P9-BYZ-589

Life Processes of Plants

LIFE PROCESSES OF PLANTS

Arthur W. Galston

**SCIENTIFIC
AMERICAN
LIBRARY**

A Division of HPHLP

New York

Library of Congress Cataloging-in-Publication Data

Galston, Arthur William, 1920–
    Life processes of plants / Arthur W. Galston.
      p. cm.
    Includes bibliographical references (p.   ) and index.
    ISBN 0-7167-5044-9
    1. Plant physiology. I. Title.
    QK711.2.G345   1994   93-31986
    581.1—dc20   CIP

ISSN 1040-3213

PRINTED IN THE UNITED STATES OF AMERICA

Scientific American Library
A division of HPHLP
New York

Distributed by W. H. Freeman and Company
41 Madison Avenue, New York, New York 10010
20 Beaumont Street, Oxford, OX1 2NQ, England

1 2 3 4 5 6 7 8 9 0 KP 9 8 7 6 5 4 3

This book is number 49 in a series.

Contents

60.609

THIS BOOK IS DEDICATED TO THREE MASTER
TEACHERS WHO INFLUENCED ME GREATLY:

Loren C. Petry of Cornell University

Harry J. Fuller of the University of Illinois

James F. Bonner of Caltech

Preface

During my half century as a plant physiologist, I have lectured frequently to students and lay audiences alike on biological subjects involving both plants and animals. With only occasional exceptions, I have noted that my listeners tend to "turn off" at first when the discussion moves to plants. I am sure this disengagement was traceable to a perception that plants just don't seem to do anything.

But a little contact with modern plant biology can turn boredom into fascination and avoidance into eager study. I discovered this myself at Cornell University in 1936 as a young student bound for veterinary school, who happened by chance to take a freshman botany course merely to round out an otherwise full course schedule. I have been forever grateful for the serendipity that brought me into contact with Loren Petry, a bearded, pipe-smoking paleobotanist whose elegant lectures enthralled me and whose Saturday evening informal discussions on any subject in the world changed the pattern of my life. Thanks to him, I began to see plants as the responsive and complex organisms they are. We are blinded to their sensitivity and complexity by their differences from the animal world, yet these very differences are the source of their fascination.

Consider for a moment how "clever" plants are, and how they solve many of the major problems of life smoothly, without noise, waste, or pollution. Their conversion of light into chemical energy for basic nutrition is still unequaled by the best human engineers, and it requires only the simplest of building blocks, namely water and carbon dioxide. Unsurpassed as organic chemists, they are also ingenious engineers. They can raise water to a height of 300 feet, using only the energy of the sun. They package their dormant seeds and buds with insulation and waterproofing agents that would make any designer envious. Although fixed in one location by their roots, they move crisply when necessary to accomplish vital tasks, without the muscle and nervous system we think of as necessary for motion. Among the most unexpected characteristics of plants is their time-telling ability. Plants can measure the time of day within a few minutes, and they can tell the date with enough accuracy to forestall injury or death from seasonal challenges.

Plants also have a busy social life, interacting with other creatures in a multitude of ways. They ward off insect predators by physical devices and defensive chemical weapons, but attract helpful organisms by variations of the same themes. They have learned to cooperate with bacteria and fungi, and even use certain bacteria to carry out genetic engineering in their cells. Certainly such extraordinary creatures are worthy of our diligent study and emulation!

One of the most remarkable powers of plants is their ability to regenerate organs; even an ordinary single cell can build the entire organism. Because of their regenerative abilities and the existence of bacteria able to transform their cells, plants have become the center of a new biotechnology industry. Humankind is developing a new power to genetically engineer plants that will revolutionize both agriculture and our own understanding of these organisms.

My undergraduate years at Cornell offered a glimpse of the remarkable abilities of plants that widened as I continued my studies. Harry Fuller, my graduate adviser at the University of Illinois, opened my eyes to the excitement of plant developmental biology, especially photoperiodism and plant hormones. In his wisdom, he had me research the physiology of flower formation in a then new agricultural plant, the soybean. My thesis title was corrupted by my fellow graduate students to "The Sex Life of the Soybean"; it didn't sell well, but it got me my graduate degree. Later, as a postdoctoral fellow in the laboratory of James Bonner at Caltech, I caught research fever and learned how independent workers in a common laboratory could stimulate one another by word and deed. It has been a happy journey since then. In this book, I would like to share with readers some of the joy that the study of how plants work has brought to me.

Academic life at most universities involves both research and teaching, since the understanding that flows from research is supposed to increase one's expository skills. Despite this credo, most universities find themselves unable to support financially the research of their faculty members. My scientific career would have been impossible without periodic grants and fellowships from private and public foundations. Thanks are due to the Guggenheim, Fulbright, and National Science Foundations for fellowships permitting study in Stockholm (1950–51), Canberra (1960–61), and London (1968), respectively. I am also deeply grateful to Yale University for its enlightened Triennial Leave of Absence policy, which facilitated shorter periods of study at Stanford; Cambridge, U.K.; Rehovot, Israel; and Tokyo. The National Science Foundation has supported my research throughout my academic career, from the year of the Foundation's creation to the year before my retirement. The National Institutes of Health, American Cancer Society, Herman Frasch Foundation, U.S.-Israel Binational Agricultural Research and Development Fund, and the Albert Nerken Fund have also aided in some years. The National Aeronautics and Space Administration deserves special thanks for continuing its support into my formal "retirement," which is turning out to be very active, indeed. I also thank Yale University for its enlightened policy permitting retirees to continue an active scholarly life.

Writing a book on science for the intelligent lay public is difficult at best, and those of us who have contributed mainly to scientific journals for many years are doubly handicapped by our enforced editorial conformity to a generally deadening style. I thank my editor, Susan Moran, for helping me to rediscover "peoplespeak," a form of the English language not widely used by the scientific professional, but useful in communicating ideas directly to lay people. I am also indebted to Travis Amos, whose skill in locating appropriate photographs has helped make this book so attractive. Janet Tannenbaum has aided greatly in the final stages, and others at the editorial offices of the Scientific American Library have also made the cooperative production of this book mostly a pleasure. The sympathetic understanding of family and friends has of course been absolutely essential to the sustained effort required to produce this book.

Arthur W. Galston
New Haven, Connecticut
September 1993

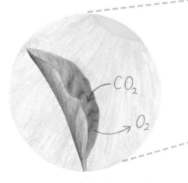

So far as we now know, all life in the universe exists at or near the surface of planet earth. Although living forms of some sort may eventually be discovered elsewhere in the cosmos, we have no hard evidence to support their existence. Certainly neither our probes of the surfaces of Mars and other planets nor our scanning of the skies for alien radio signals have yielded any positive result. We thus have to believe, at least for the present, that the life we see is the only life there is, and that we live in a unique place in the universe.

What makes the earth's surface so suitable for the origin and persistence of life? Three interacting factors are probably critical: a temperature range permitting the existence of liquid water, an abundance of visible light energy without too much harmful ultraviolet or infrared radiation, and an unusual atmospheric chemistry, including oxygen. All three conditions essential to modern terrestrial life forms have arisen through a single unique activity of green plants: their use of light energy absorbed from the sun to split apart ("photolyze") ordinarily stable molecules of water, H_2O, thereby liberating gaseous oxygen, O_2. Since each molecule of water contains only one atom of oxygen, but each molecule of oxygen gas contains two atoms of oxygen, two molecules of water must be photolyzed for each molecule of oxygen liberated.

The Importance
of Being Green

The photolysis of a water molecule is one of the steps in photosynthesis, the process by which the plant taps the energy of the sun to synthesize its own supply of food. When the water molecule is split, the hydrogen is not released as hydrogen gas, H_2, but rather as the constituent protons and electrons of which each hydrogen atom is composed. Within the leaf cell, the protons and electrons from hydrogen are ultimately used to transform molecules that have been formed from carbon dioxide, a gas that the plant absorbs from the atmosphere. By combining the constituents of hydrogen atoms with carbon dioxide and other components, the plant synthesizes sugars used to store energy and construct the plant body.

Plants have transformed the atmosphere on which life depends by releasing oxygen on the one hand and absorbing carbon dioxide on the other. The primitive atmosphere of the earth is believed to have consisted largely of hydrogen, methane, ammonia, and water, along with some carbon dioxide. This mixture would be toxic to all but a few very simple forms of life as we know it. By contrast, our present atmosphere is noteworthy for its abundance of free oxygen (about 21%). Abundant free oxygen is an absolute requirement for almost all higher life forms, and virtually all of the oxygen in our atmosphere is believed to have arisen from the simple splitting of water molecules in the plant leaf.

Sunlight, water from the soil, and carbon dioxide from the air interact in higher green plants, producing sugar and oxygen. The photolysis of a water molecule is a central event in this process of photosynthesis: the protons and electrons of the hydrogen atoms are used to convert radiant energy to a chemical form.

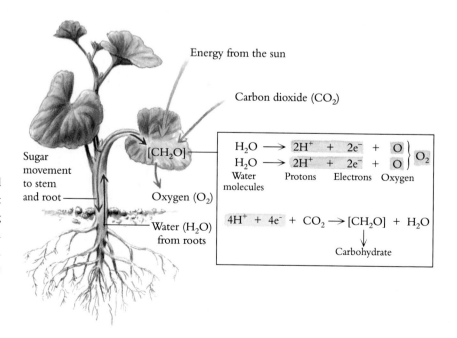

Energy from the sun

Carbon dioxide (CO_2)

Sugar movement to stem and root

[CH$_2$O]

Oxygen (O_2)

Water (H_2O) from roots

$$H_2O \longrightarrow 2H^+ + 2e^- + O$$
$$H_2O \longrightarrow 2H^+ + 2e^- + O$$ $\Big\} O_2$

Water molecules   Protons   Electrons   Oxygen

$$4H^+ + 4e^- + CO_2 \longrightarrow [CH_2O] + H_2O$$
Carbohydrate

Some of this oxygen, O_2, is converted to ozone, O_3, in the stratosphere through the action of ultraviolet light. Stratospheric ozone protects life on the earth's surface by absorbing much of the harmful ultraviolet radiation coming from the sun. The planet Mars offers an example of an environment hostile to life, where atmospheric ozone is absent. The action of ultraviolet radiation has transformed the mineral surface of that planet into a corrosive oxidant that would destroy molecules of living matter.

Most of the gaseous carbon dioxide in the ancient atmosphere was converted to sugar in growing plants. Thus, the earth's modern atmosphere also differs from its ancient predecessor in its extremely low content of carbon dioxide (about 0.033%). This alteration in atmospheric chemistry, by preventing the excessive heating of the earth through carbon dioxide–mediated entrapment of infrared radiation, has indirectly made possible the presence of liquid water.

Since a watery fluid is the basic medium filling all living cells, a temperature range permitting water to exist in liquid form is probably the most important condition for the existence of life on earth. When cellular water freezes, it becomes biologically unavailable, and almost all life processes slow to a halt. When it becomes too hot, it causes proteins and other cell components to become "denatured." Boiling an egg to coagulate the white is an example of heat denaturation; in the "native," translucent state, the protein molecules of egg white are coiled into precise configurations, but when boiled, they open up and lose their shapes.

Life enjoys a limited range on our planet, circumscribed mainly by temperature. The earth is a roughly spherical body with an equatorial radius of almost 6400 km, covered by a thin crust of solidified granitic material averaging only about 5 to 8 km in thickness under the oceans and 25 to 50 km where there are continents. This fragile shell is covered by water for about four-fifths of its surface area and four-fifths of its average height. In the waters of the earth and on the thin solid crust protruding above are found all the various forms of life. But the terrestrial zone where liquid water may exist is very narrow, for at altitudes only several kilometers above the average height of the earth's surface, temperatures are low enough to freeze essentially all water, and at depths several kilometers below the earth's surface, temperatures are so high that water can boil.

No life forms can grow and flourish when frozen, but some cells that have developed in more permissive temperatures can remain alive, although quiescent, in the frozen state. Sometimes, such cells can resume their normal activities when they are thawed and kept at warmer temperatures. In the laboratory, seeds, spores, sperms, and even some animal embryos can be stored in the frozen state for years and then revived essentially unharmed. The main trick is to prevent the formation of large crystals of ice, whose growth through the cell destroys essential structures such as the surrounding membrane. Interestingly, it has recently been found that DNA, the molecule carrying the genetic information of the cell, is not necessarily disrupted by freezing.

At the other end of the temperature scale, the buildup of heat below the earth's crust, due to radioactive decay, makes the existence of any stable life forms unlikely very far below the surface. For each kilometer that we descend beneath the earth's surface, the temperature increases by about 30 °C. Thus, at roughly 3 km down, we reach the boiling point of water, beyond which only a few unconventional life forms can exist. Recently bacteria isolated from boiling hot springs have been shown to survive temperatures above 100 °C for extended periods. These unique cells seem to contain special chemical substances that stabilize cellular proteins and other cell components against heat denaturation. No organism higher than a bacterium is capable of surviving near the boiling point of water.

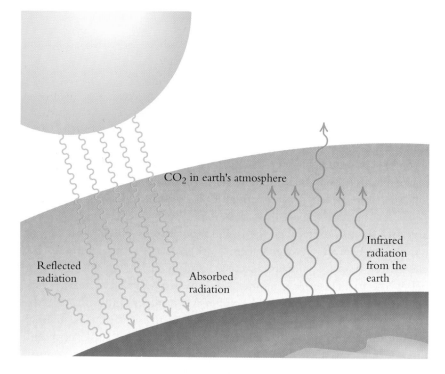

Carbon dioxide is transparent to visible light but opaque to longer-wavelength, infrared radiation. CO_2 in the earth's atmosphere thus warms the earth by trapping the infrared radiation emitted by the sun-warmed earth.

CO_2 in earth's atmosphere

Reflected radiation

Absorbed radiation

Infrared radiation from the earth

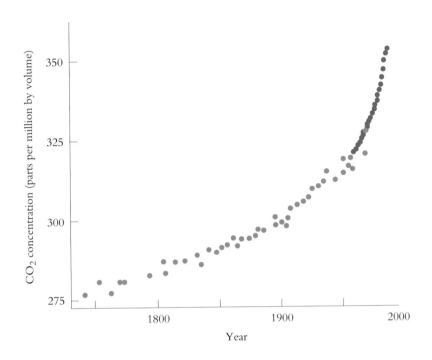

The carbon dioxide content of the earth's atmosphere has been rising steadily for the last two centuries. The green circles represent data from an ice core at Siple Station, Antarctica; violet circles represent data recorded at Mauna Loa, Hawaii.

For most life to survive, the planet must not become so hot that water vaporizes. The sun's rays warm the earth, which then reradiates some of this absorbed heat energy into space in the form of infrared radiation. Carbon dioxide, which is transparent to visible light, absorbs radiation at infrared wavelengths. Thus the level of carbon dioxide in the atmosphere must not become too high; otherwise, this "greenhouse gas" will trap heat energy in the atmosphere. The very high level of carbon dioxide in the atmosphere of Venus, for example, may be responsible for that planet's heat and lifeless aridity. Fortunately, the green plants on earth absorb considerable carbon dioxide for use in photosynthesis, and at present the level of this gas in our atmosphere is somewhat stable, although it has risen steadily since the beginning of the industrial revolution, when the combustion of fossil fuels started to pour additional carbon dioxide into the air. The uptake of carbon dioxide for photosynthesis can just about balance the release of carbon dioxide from burning, respiration, and the weathering of rocks, provided that we do not destroy too many of our forests, especially those in the tropics, which have been called the "lungs" of our planet.

Plants and Sunlight

The sun is essentially a thermonuclear device powered by a reaction in which four hydrogen atoms are fused to form one helium atom. In this process, a small amount of mass disappears, giving rise to large amounts of energy according to Einstein's well-known equation, $E = mc^2$. Some of this energy strikes the surface of the earth in the form of visible light. The uniqueness of green plants lies in their ability to absorb and then convert the energy in selected wavelengths of light into a chemical form.

Light is usually depicted as a wavelike motion, described by its wavelength and frequency. Wavelength (symbolized by the Greek letter λ) is the distance between successive peaks of a wave; it is measured in nanometers (10^{-9} meters), abbreviated nm. Frequency (ν) is a measure of the number of wave peaks passing a given location each second. Visible light spans wavelengths from about 400 to 700 nm; we perceive the different wavelengths as different colors.

Although light has the properties of a moving wave, it is also known to occur in discrete packets called photons, each with a fixed amount of energy called a quantum. The energy of any quantum (E) is proportional to the frequency ν at which the wave oscillates and is thus inversely proportional to the wavelength λ, since $\lambda \times \nu$ equals the speed of light, a constant. A quantum of blue light of wavelength 400 nm thus has a higher frequency and is more energetic than a quantum of red light of wavelength 700 nm. If captured by a plant, a quantum of sufficient energy can be put to work driving chemical reactions such as those by which a plant forms sugars during photosynthesis.

In photosynthesis, radiant energy is captured by special molecules called pigments, contained in elaborately structured devices, the chloroplasts of higher plants and the chromatophores of certain bacteria. The electrons in these pigments occur at certain energy levels that permit them to capture quanta of light at specific frequencies or colors. These frequencies represent an amount of energy sufficient to raise the electrons from their ordinary, or "ground state," energy level to an "excited state." The electrons that become excited are the outermost electrons of an atom, the electrons that are shared by neighboring atoms to form the chemical bonds that hold together the atoms of a molecule. The excitation of its electrons gives the bond a higher-than-normal energy, which can be transferred to the bonds of other molecules. It is at such locations of high energy within the molecule that bonds can be broken or rearranged. The excited molecules can thus use their extra energy to drive chemical reactions that would not otherwise proceed readily.

The main light-capturing pigment of most plants is chlorophyll, a green substance capable of absorbing quanta of both red and blue light. (It appears green to our eyes because it does not absorb green light, but rather transmits it. The transmitted green light is the light that we see.) After a chlorophyll molecule absorbs a red or blue quantum, becoming "excited," it contains the energy of the quantum, now stored in the form of an altered chemical bond. Other chloroplast or chromatophore pigments, such as the yellowish carotenoids or bluish phycobilins, can also capture light energy, which they pass on to chlorophyll at a "reaction center" in the chloroplast. There, the stored energy is used to drive reactions in the cell that lead eventually to the synthesis of stable compounds. The intimate details of this intriguing process will be discussed in Chapter 1.

Photosynthesis is an example of energy transduction: the conversion of one kind of energy into another. The sugars synthesized by the plant contain the energy of the original absorbed light in chemical form. Whether they remain as separate sugar molecules or are linked together in chains to form complex carbohydrates, they function as the plant's food, and will yield up their energy when combined with oxygen during respiration. Thus photosynthesis and respiration are reciprocal processes: photosynthesis releases oxygen and stores energy, while respiration absorbs oxygen and releases energy.

The photosynthesis of sugars and other foods from carbon dioxide and water provides an energy source for the cellular activities of other living organisms. When we eat a plant, we tap the solar energy it has captured and stored; similarly, when we eat an animal that has eaten a plant, we are indirectly harvesting the energy of the sun.

Recent deep-sea explorations have revealed complex communities of higher organisms around hot volcanic vents in the ocean floor. These creatures feed on heat-tolerant bacteria that literally "chew iron and spit rust"; the bacteria obtain their energy mainly by oxidizing two unique "foodstuffs," the elements iron and sulfur, which are abundant in the

White light contains all the colors of the spectrum, which can be displayed when light is diffracted by a prism. Leaves look green because they transmit green light and neighboring frequencies of white light while absorbing heavily in the red and blue parts of the spectrum.

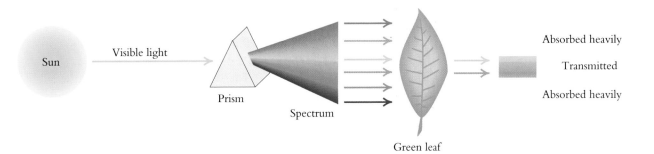

Sun   Visible light   Prism   Spectrum   Green leaf   Absorbed heavily   Transmitted   Absorbed heavily

Long chains of cells are formed by chemosynthetic bacteria that grow near hot vents on the ocean floor. The bacteria store energy by oxidizing iron and sulfur and are the base of a complex living community on the ocean floor. The stalks increase the surface area available for absorbing essential raw materials.

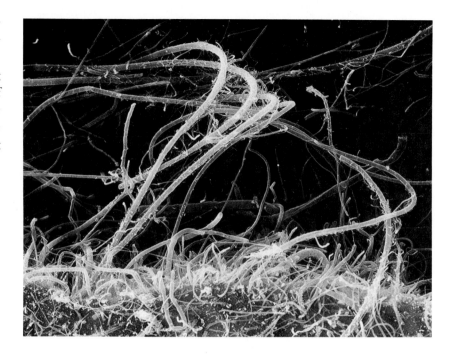

magma issuing forth from the earth's interior. This mode of obtaining nutritional energy from inorganic or organic chemicals is called chemosynthesis. It is limited to a few "bizarre" groups of microorganisms. In general, we may thus view green plants as the main support for other forms of life on earth.

Green Design: A Look at the Plant Body

The plant body has a simple and logical design, efficient for collecting light. For photosynthesis to be effective, the plant must present to the sun a light-absorbing organ, the leaf, oriented at a proper angle to the incident radiation. Each leaf must have a large surface area to maximize its absorption of light energy, a limited thickness to minimize the attenuation of light reaching the bottom cell layer, and a means of adjusting to the changing position of the sun in the sky. Carbon dioxide must be able to enter and spread through the leaf, and oxygen must be able to exit. Both gases are carried in extensively branched gas channels within the leaf that connect to the outside world through pores in the leaf surface called stomata.

Individual leaves are positioned around a central axis, or stem, in such a way that leaves fall in the shade of higher leaves as little as possible. Leaves are inserted into the stem in a fixed pattern that is genetically determined. For example, in some plants, two leaves are found opposite each other at nodes, and the orientation at successive nodes is displaced by 90°. Other plants will show a fixed spiral arrangement of leaves around the stem. The number of leaves intercepted during each revolution around the stem as one moves up the spiral is usually represented by a fraction called the phyllotaxy. Thus, opposite-leaved plants have a phyllotaxy of 1/2, and 2/5 phyllotaxy means that five leaves are intercepted during two revolutions.

Although plants are generally considered to be immobile, a leaf that falls into the shade despite its positioning away from other leaves may be able to move back into the sun. Leaves are able to detect the changing

A tobacco plant has 2/5 phyllotaxy: leaf 1 arises exactly over leaf 6, leaf 2 exactly over leaf 7, and leaf 8 exactly over leaf 3. Thus, five leaves are intercepted during two revolutions around the stem. This arrangement optimizes light falling on individual leaves.

orientation of the sun during the day and can avoid excessive shading by adjusting the position of the leaf blade through changes in special cells in an organ at the leaf base called the pulvinus. This is but one example among a variety of plant movements that will be examined in Chapter 4.

The green plant is an organic chemist par excellence. To supply its needs, it synthesizes not only the sugars that are the immediate product of photosynthesis but many more complex molecules, from basic nutritional requirements and growth regulatory hormones to a woody skeleton, corky outer bark, and waxy protective coating of leaves and fruits. So far as we know, all these molecules are synthesized from the simplest of starting materials: carbon dioxide from the air, water and mineral elements from the soil, and light energy from the sun.

The green plant requires a total of 17 elements of the periodic table; of these, carbon, hydrogen, oxygen, and nitrogen come from the atmosphere and from water, and the other 13 come from the weathering of minerals in the soil. The minerals are both structural and catalytic components of the plant cell; for example, magnesium is part of the chlorophyll molecule, phosphorus is a component of the nucleic acids that make up the genetic material, and iron and copper are essential parts of some enzymes. Although 79 percent of the earth's atmosphere is nitrogen gas, N_2, plants are unable to make use of the element in its gaseous form. The nitrogen must first be converted to ammonia, NH_3, by some free-living microorganisms and certain symbiotic associations of microbes and higher plants. How plants and some microbes are able to function together as almost single organisms will be discussed in Chapter 7.

Below the soil surface, highly branched roots tap the water and minerals of the soil, and anchor the plant firmly. To ensure an adequate supply of water and nutrients to the leaf, the stem contains a vascular system consisting of open tubes running all the way up from the root. The tubes of the vascular system extend from just above the root tip to just below the stem apex and into each leaf, branching so profusely that no individual cell is far from contact with the vasculature.

Some specialized cells of this system conduct water and dissolved materials drawn from the soil up to the leaves. Most of the absorbed water replaces that which has been evaporated from the leaves by the sun's warmth, although some is used in photosynthesis. The water-conducting cells are either long, thin tracheids or short, fat cells stacked to form vessels; cells of both types form the xylem. These cells are programmed to die shortly after they reach maturity, leaving behind long, empty, thick-walled tubes that are well adapted for the transport of water and dissolved mineral nutrients from the soil. Outside the xylem is another type of

Facing page The parts of a higher green plant arise from meristems at the stem apex, which produces stem, leaf, and flower tissue; at the root apex, which produces the root system; and at the cambium, which increases girth. A vascular system connects all the parts. Sections through the specific regions are shown on the right.

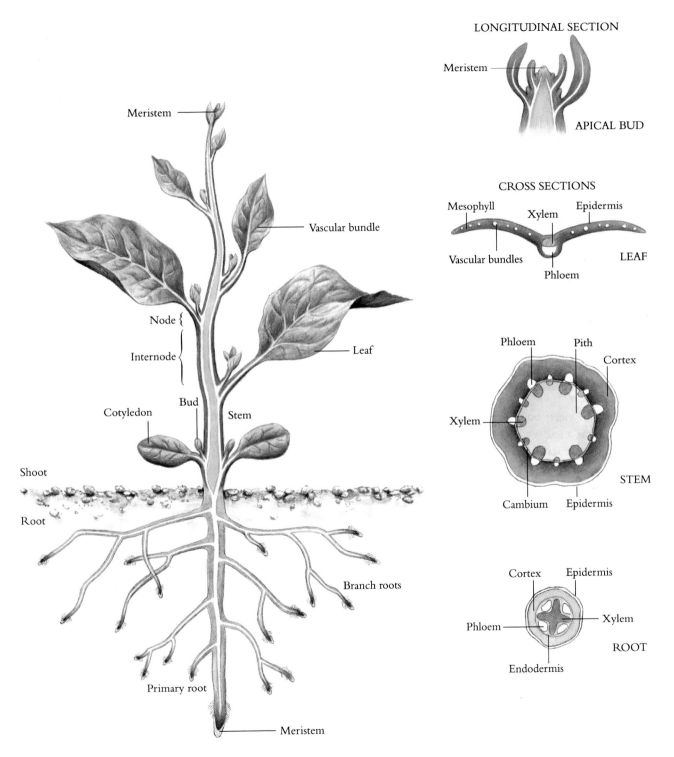

LONGITUDINAL SECTION

Meristem

APICAL BUD

CROSS SECTIONS

Mesophyll   Xylem   Epidermis

Vascular bundles   Phloem   LEAF

Meristem

Vascular bundle

Node {

Internode {

Leaf

Cotyledon   Bud   Stem

Shoot

Root

Branch roots

Primary root

Meristem

Phloem   Pith   Cortex

Xylem

Cambium   Epidermis   STEM

Cortex   Epidermis

Phloem   Xylem

Endodermis   ROOT

Plants growing in the Sonoran desert in Mexico are under severe stress from heat and lack of water most of the time. Their various adaptations increase their chances of survival in this harsh environment. For example, the spines of cacti represent reductions of large leaf surfaces that would evaporate considerable water. In many cacti, photosynthesis is carried out in cylindrical wax-coated stems.

molecules may exist per cell, yet small changes in this number can greatly affect whether a cell divides, elongates, goes into dormancy, reacts to light or gravity, or dies.

We now recognize five different types of plant hormones that regulate growth, development, and certain aspects of plant behavior. As we shall see in Chapter 3, these compounds have been isolated, chemically identified, and studied rigorously by quantitative physical, chemical, and physiological techniques. In addition to these well-recognized substances, we have good evidence to suspect the existence of other molecules that control flowering, senescence, reactions to wounding and microbial invasion, and the rapid movements of some tendrils, stamens, and leaves. Our knowledge of how plants regulate their activities by hormones is thus only fragmentary, and the subject is still under active investigation.

Anticipating the Future

Many trees protect themselves against freezing winters by shedding their leaves and other succulent, easily freezable parts well in advance of the first frost. At the same time, they develop well-protected winter buds, containing woolly insulation hairs and surface scales that protect the delicate growing point. To make these preparations efficiently, trees anticipate the approximate date of the first frost. Plants are able to predict the seasonal changes ahead by measuring signals from the environment such as the daily change in length of the interval between sunrise and sunset.

It takes the earth about 24 hours to rotate around its tilted axis and about 365 days to revolve around the sun. Any point on the earth's surface is thus exposed to a pattern of solar radiation that changes hour by hour and day by day. The daily cycles of light and dark, accompanied by temperature changes, and the larger seasonal changes in temperature and precipitation are consequences of these rotations. Organisms like plants that depend on the absorption of light energy and that must live through great changes of temperature and water availability have evolved sensitive mechanisms to cope with these changes.

Plants in the earth's temperate zones have met these challenges successfully by combining two physiological mechanisms, a biological "clock" that measures the passage of a 24-hour period and a pigment called phytochrome that is able to perceive the beginning and end of the daily light period. Since the length of this daily light period varies with the season, the plant must integrate the two signals to determine the present date and thus the probable future course of climatic events. Thus, a progressively decreasing daylength can be used to predict the onset of winter, and the exact length of day to predict the approximate time until the first killing frost. The details of such plant "computing" are presented in Chapter 2.

Regeneration from Individual Cells: Totipotency and Immortality

Despite the presence of some woody cells that give the plant body its basic form and strength, the plant is a fragile organism, subject to injury from wind, physical impact, the grazing of animals, and invasion by microorganisms. To compensate for this fragility, the plant has not only developed various mechanisms for coping with stress, explored in Chapter 5, but it has also developed a potent capacity to regenerate its body, discussed in

Chapter 6. Thus, the plant "tries" to avoid injury to its parts, but if it fails, every injured or discarded organ is potentially replaceable.

The remarkable regenerative capacity of plants is displayed most vividly by single cells. Like all multicellular organisms, the higher green plant originates from a single cell, usually the fertilized egg. Because this special cell is able to divide successively to produce the entire organism, it is said to be "totipotent," or all powerful. About 60 years ago, it was discovered that bits of tissue from almost any part of the plant could be excised and grown in cultures made by placing the tissue in flasks containing artificial media with all essential nutrients. Usually, the new cells were not organized into roots, stems, or leaves, but produced a somewhat formless mass called a callus. The callus could be subdivided and subcultured repeatedly without any diminution of the growth rate. There is general agreement that such cells are potentially immortal; that is, they will continue to grow as long as they are provided with fresh medium from time to time. On the plant, however, such cells may have a finite life span, suggesting that some regulatory substances in the plant may program their death.

Occasionally, if the appropriate growth hormones are supplied, roots and buds form on previously undifferentiated callus masses. When these bits of tissue are properly transplanted, they can give rise to entire plants. Because plants produced in this manner are presumably identical genetically, this technique, called cloning, has become an important means of propagating desirable lines of plants used in agricultural biotechnology. About 30 years ago, thanks to advances in our knowledge of plant growth hormones, it became possible to cultivate even a single plant cell in an artificial medium. The isolated cell can be induced to divide to produce a callus mass, which then differentiates to grow into an entire plant. This important experiment shows that every plant cell not only has the genetic information for constructing the entire organism, but is actually totipotent. Thus, to establish a new genetic line, all one has to do is to clone a single cell of the desired type. After the recent development of genetic engineering techniques for transforming individual plant cells, this procedure has become the basis for an important new industry. We shall examine many aspects of these fascinating new procedures in Chapter 8.

Certain plant tumors, especially those belonging to the crown gall group, consist of cells that appear to be autonomous, even on the plant. These tumors may appear as single isolated masses on the plant or as systemic outgrowths at many points on the plant body. If the cells of such tumors are placed into tissue culture, they grow very rapidly and seem to be potentially immortal, although they rarely differentiate any formed organs. Remarkably, these cells continue growing without the addition of

A tobacco callus has given rise to flower buds (*left*) and leaf buds (*right*). Entire plants can be regenerated from such calli.

any of the plant hormones needed by "normal" growing cells. Experimenters reasoned that the ability to grow independently of growth hormones produced elsewhere in the plant accounts for the tumorous nature of such cells.

Later, it was found that crown gall tumors contain certain cellular components of a microorganism called *Agrobacterium tumefaciens*. When virulent cells of this bacterium infect a normal cell, they transfer to the cell a small plasmid, a ring-shaped body containing some of the DNA that constitutes the genetic material of the bacterial cell. Part of this plasmid DNA is then inserted into the DNA of the plant cell; when the plant cell divides, the inserted DNA is reproduced along with the normal genetic material. The genes in the foreign DNA control the synthesis of the plant hormones and other processes that make the receptor cell tumorous. Thus, infection by *Agrobacterium* leads to the transfer of the bacterium's genetic material to the plant cell, a kind of natural genetic engineering. This is but one example of a useful cooperation between plants and microbes, analyzed in Chapter 7. Genetic engineers can put this remarkable ability to use by deliberately inserting altered genes into the plasmid for transfer to a receptor plant cell, using techniques to be described in Chapter 8. Such genetic transformation of a single plant cell, followed by cultivation and cloning on artificial media, has already led to the development of new, agriculturally important plant strains.

Understanding how plants work thus requires some mastery of a number of scientific disciplines, certainly including the biological subdisciplines of anatomy, physiology, and genetics, but also including physics, chemistry, and astronomy. We shall be making use of all of these disciplines, and more, as we proceed through this book.

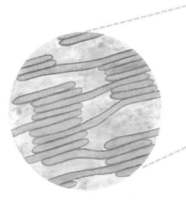

We have not always understood that plants synthesize their own food. The ancients believed that plants derived their food from the soil, and that the root system was a kind of diffuse mouth sucking nutrition from the earth's breast. This notion was proved wrong early in the seventeenth century, when a Dutch physician named Jan Baptista van Helmont planted a 5-lb willow sapling in a container of dried soil weighing exactly 200 lb. For five years, he added nothing to the container except rainwater. When he harvested the tree, he found its weight to be 169 lb, while the dried soil now weighed 199 lb 14 oz, a small loss he attributed to experimental error. This simple experiment exploded an ancient myth, showing clearly that the major part of the weight gained by a plant as it grows does not come from the soil.

What was the source of the extra weight? Since he had added only rainwater, van Helmont concluded that water somehow became incorporated into the dry mass of the plant body. This conclusion, which must have seemed mysterious and improbable at the time, correctly accounts for part of the gain in mass. Van Helmont also concluded, this time incorrectly, that plants derived no materials at all from the soil other than water. He did not appreciate that the approximately two ounces lost by the soil included mineral elements that were in fact essential to the well-being and

ONE

Photosynthesis:
Food from Photons

growth of his tree. Many modern gardeners still talk of "feeding" plants when they add small quantities of mineral fertilizer to the soil. Strictly speaking, this designation is incorrect, since foods are defined strictly as organic materials capable of releasing energy when combined with oxygen. This process, called oxidation, is what goes on when a piece of coal is burned, or when sugar is respired ("burned") in a living cell. The green plant makes all of its own food through photosynthesis, even though it absorbs the required mineral elements from the soil.

Van Helmont was also completely unaware of the existence of gases, which were not to be discovered for another century and a half. He thus could not have appreciated the bizarre fact that much of the gain in weight of the tree could be attributed to the absorption of carbon dioxide, a minor component of the air. In fact, Stephen Hales, an English cleric often referred to as the "father of plant physiology," surmised early in the eighteenth century that plants might obtain a portion of their nutrition by drawing "air" through their leaves. His concept of air was essentially Aristotelian, in that he considered it one of the basic ingredients of which all organic matter is made. An understanding of photosynthesis awaited the chemical dissection of air that was achieved late in the eighteenth century. Over a relatively short period of time, carbon dioxide, oxygen, hydrogen, nitrogen, and other gases in air were isolated by chemists like Joseph Black, Karl Scheele, Antoine Lavoisier, and Henry Cavendish. The discovery of photosynthesis in plants both drew on and contributed to the newer understanding of gases.

The "discovery" of photosynthesis is usually attributed to the radical, nonconformist Unitarian minister Joseph Priestley, who was ultimately forced to flee England for the United States because of his expressed sympathy for the principles of the French Revolution. Living above a brewery, Priestley was impressed by the bubbling of the fermenting vats containing yeast and malted barley. He became interested in the gaseous state of matter and in the effect of gases on the survival of animals. For his experiments, he invented a mousetrap that provided uninjured mice, which he then confined in another of his inventions, a pneumatic trough that produced enclosed gas spaces sealed by water. (Pneumatic troughs are used in elementary chemistry laboratories to this day.) Priestley found that enclosed air could support the respiration of a mouse until about 20 percent of the air's volume had been displaced by the water of the trough; at this point the mouse died by suffocation. (We can understand this phenomenon easily, although it was somewhat mysterious to Priestley. The 20-percent loss of volume is created because the breathing mouse absorbs oxygen from the enclosed air, while the carbon dioxide expired by the

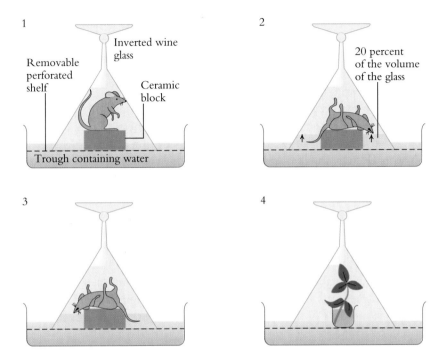

Priestley's experiment showing the opposite effects of animals and plants on enclosed atmospheres: (1) A live mouse is inserted into the enclosed atmosphere of an inverted wine glass sealed by water in a pneumatic trough.
(2) As the mouse respires, the water rises in the glass; when 20 percent of the volume of the glass has been displaced by water, the mouse dies. (3) If the dead mouse is removed and replaced by a second live mouse, the second mouse dies immediately without further change in the volume of the atmosphere in the glass. (4) A sprig of spearmint placed in an enclosed atmosphere does not die, nor does it reduce the volume of the enclosed space. In fact, it reverses the 20 percent loss caused by the mouse.

mouse dissolves in the water, causing a net loss of volume in the gas phase.) Priestley remembered that a burning candle confined to the enclosed space also snuffed out spontaneously when about 20 percent of the air had been displaced. He concluded that animals and burning objects converted pure air into impure air, and that impure air could not support animal life.

Two additional observations led to a breakthrough in Priestley's thinking. When the contents of bubbles from beer fermentation, or the "impure air" produced by mouse respiration or the burning of a candle, were substituted for pure air in the enclosed container, they could not support any respiration at all. A mouse confined to that chamber suffocated promptly without any further loss in the volume of the enclosed air. Clearly, something had been removed from the substituted air samples that was essential for mouse respiration.

Because Priestley was a great believer in the unity of living things, he expected that a sprig of spearmint confined to an enclosed space would, like a mouse, render the air "impure." To his surprise, the plant flourished in the enclosed container, growing even more vigorously than a similar unconfined plant. His great discovery, published in 1772, came when he found that air rendered impure by a confined mouse or a burning candle could be "purified" by a plant confined for some time in the same space. (As we know now, the oxygen removed by mouse or candle is restored by the plant.) Priestley's discovery led him to enunciate the general principle that animal life and vegetation live in a reciprocal relation with each other; animals make the air impure and unfit for continued existence, while plants repurify it, thus making further animal life possible. This is, of course, a statement of the reciprocal relation of photosynthesis and respiration, described in the Prologue.

While Priestley was convinced of the truth of his observations and of the generalization that flowed from them, he admitted to being troubled by the inconsistency of his results. Most of the time, the plants were effective in purifying the impure air, but sometimes they failed. He never found out why, but we can surmise that the illumination varied from experiment to experiment, not only because of changes in the weather or the time of day, but also because Priestley may have varied the location of experiments within his laboratory. That light played a definitive role in Priestley's successful experiments was established in 1779 by Jan Ingenhousz, a Dutch physician who was spending some time in England repeating and extending Priestley's observations. He found that under brilliant illumination plants could purify impure air within several hours rather than several days, as described by Priestley. Ingenhousz's experiments were in turn repeated and extended by Jean Senebier, a Swiss pastor from Geneva working at about the same time. Remembering Priestley's observation that spearmint grew better in impure air than in open air, he found that impure air also increased the plant's ability to purify air in the presence of light. Air enriched in an unknown substance called "fixed air" had the same effect.

At about the same time, the labors of Antoine Lavoisier and other pneumochemists were providing information about the nature of "fixed air," "impure air," and "pure air." Impure air differed from pure air in two important respects: it lacked oxygen, which was required to support respiration or burning, and it was enriched in carbon dioxide, which was identical with fixed air. Having this new knowledge in mind, Ingenhousz seized upon Senebier's observations to establish that plants retained weight from the carbon dioxide in the fixed air they absorbed. This finding vindicated the much earlier prediction of Stephen Hales that air played a role in plant nutrition.

It was not long before the use of the new quantitative chemistry permitted the Swiss botanist Nicolas de Saussure to discover that the gain in dry weight by plants far exceeded the weight of the carbon dioxide absorbed. Since the plants were growing in pure water, he reasoned, in a throwback to van Helmont, that water was incorporated into the dry matter formed during the process of air purification. The final great insight came in 1845, not from chemistry but from physics, through the efforts of Robert Mayer, a German surgeon who had earlier enunciated the Law of the Conservation of Energy ("Energy can be neither created nor destroyed"). Mayer showed that combustion of the organic matter produced in photosynthesis liberates an amount of energy equivalent to the light energy absorbed by the plant. Thus, by a century and a half ago, scientists had come to understand the basic nature of photosynthesis: it is a mechanism for converting the radiant energy of the sun into a stored chemical form.

Chlorophyll and the Chloroplast

In the modern era of research in photosynthesis, investigators have attempted to dissect the process into its component parts, to understand each part of the process in intimate detail, and to reassemble the parts into a comprehensible whole. Two basic approaches to this quest have been valuable. The observation of the cell and its components has revealed the importance of the chloroplast in photosynthesis; then, the use of the analytical techniques of physics and chemistry has provided a cornucopia of details, which we are only now assembling into a single picture.

One can learn much about the mechanisms involved in the photosynthetic action of light by simple physical means, without turning to complicated chemical analyses. About 60 years ago, Robert Emerson of Caltech and William Arnold of the Oak Ridge National Laboratory found, by means of flashing light experiments, that photosynthesis could be separated into two groups of chemical reactions: one group that can take place in the dark and one group that requires light. Emerson and Arnold directed a light beam of measured energy onto a plant through a rotating disc divided into adjustable transparent and opaque sectors; by adjusting the size of each sector and the speed of the disc's rotation they provided light and dark intervals of varying durations. This experiment tested how different intervals of darkness affected the rate of subsequent photosynthesis, as measured by the amount of oxygen liberated. The experiment produced a surprising result: each standardized flash of light produced its greatest photosynthetic yield when it was preceded by a somewhat longer dark period. Emerson and Arnold interpreted this result to mean that slower chemical reactions not requiring light prepared the way for the rapid action of light.

We now associate the slower "dark reactions" with various steps in the initial retention and transformation of carbon dioxide. Before the carbon dioxide entering a leaf can be acted on by light, it must first be "fixed" by combination with another chemical substance. The newly formed compound then goes through several light-independent chemical modifications. The rapid light reaction yields its own, separate products, which interact with the modified, fixed form of carbon dioxide to yield the sugars that are the final products of photosynthesis. The various dark reactions are not unique to photosynthesis but are carried out by the same general chemical machinery of the cell used in energy metabolism. By contrast, the light reaction, which comprises various steps leading to the cleavage of a water molecule, is absolutely unique to photosynthesis. During this reaction, light energy is transduced into a storable chemical form that can be used later as needed by the cell.

Light, to be effective, must first be absorbed by a pigment, a substance that by definition strongly absorbs visible light. Most pigments absorb light of only certain wavelengths and transmit light of all other wavelengths. Each pigment can be considered to have its own "fingerprint" consisting of the wavelengths it absorbs and the strength of absorption. This characteristic of pigments gave scientists a way to identify the pigment responsible for photosynthesis. Even when the pigment was completely unknown, its "fingerprint" could be obtained by finding those wavelengths of light that cause the plant to produce the greatest photosynthetic activity. When the relative photosynthetic activity is plotted against wavelength, the result is an *action spectrum*. Ideally, the action spectrum should match the relative absorption of different wavelengths by the effective pigment (the *absorption spectrum*). In practice, exact comparisons between these two types of spectra are difficult to make for several reasons: other, inactive pigments are present whose absorption overlaps that of the effective pigment; light is scattered differentially by cellular components; and light energy is transferred between different pigment systems.

Nevertheless, some knowledge about reactions caused by light can always be gained by comparing action spectra to absorption spectra.

The visible spectrum is usually considered to extend from wavelengths of about 400 nm in the blue to about 700 nm in the red, although some individuals can see down to about 380 nm in the indigo and up to about 750 nm in the far-red region. The first step in constructing an action spectrum is to break up white light into its component wavelengths by a prism, diffraction grating, or set of interference filters. The positions of the various wavelengths and their relative intensities are then established by simple physical methods. Each separate wavelength is allowed to fall on a leaf. Measurements of energy absorption and oxygen evolution constitute the data for calculating the action spectrum.

The action spectrum for photosynthesis of a typical green leaf matches very well the combined absorption spectra of chlorophyll and associated carotenoids. The details of the curve, with major peaks in the red and blue regions of the spectrum and a trough in the green, make it very likely that these pigments, and probably no others, are the primary light absorbers in photosynthesis. In red and brown algae and in cyanobacteria, still other pigments contribute to photosynthetic light absorption.

Chlorophyll and its associated pigments reside in the chloroplast, an elaborately structured entity within the cell. The typical modern chloroplast of higher plants is a wafer-shaped body about 5 micrometers long by about 2 micrometers wide and deep. (A micrometer, abbreviated μm, is a millionth of a meter.) The number of chloroplasts in a cell varies from 1 in some algae to perhaps 50 to 100 in an average leaf cell. The chloroplasts are surrounded by cytoplasm, the membrane-bounded liquid part of the cell. The cytoplasm surrounds all of the distinct entities called organelles, such as the chloroplasts and DNA-containing nucleus, that perform the cell's vital functions.

The leaf consists mainly of a flattened blade of photosynthetic cells containing chloroplasts, surrounded by a protective epidermal layer perforated by stomatal pores, and connected to the stem by a leaf stalk, or petiole. Through the petiole runs a major vein connected to the vascular system of the stem and root; this vein subdivides in the leaf blade so that almost every cell is in direct contact with it. The photosynthetic leaf blade usually consists of several layers of mesophyll cells, including one or more palisade layers made up of closely appressed columnar cells and a more open, spongy layer near the stomata that is made up of irregular cells fitting together loosely, leaving many open air spaces that connect with the stomatal pores.

Actively dividing cells at the stem apex, which give rise to leaves, do not contain green chloroplasts, but harbor instead smaller unpigmented

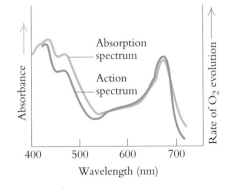

The action spectrum for photosynthesis corresponds broadly with the absorption spectrum of a green leaf.

A cross section of a leaf shows palisade and spongy cell layers sandwiched between two epidermal layers containing stomata.

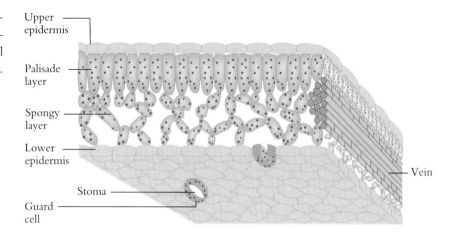

Upper epidermis

Palisade layer

Spongy layer

Lower epidermis

Stoma

Guard cell

Vein

proplastids, the form in which chloroplasts presumably replicate. Proplastids grow in size by importing materials from the cytoplasm; eventually, they synthesize chlorophyll, change their internal structure, and become true chloroplasts. In flowering plants, proplastids require light to complete their structural development. Part of this required light is absorbed by protochlorophyll, a chlorophyll-like substance formed in the dark and transformed to chlorophyll by the light it absorbs, and part of the light is absorbed by phytochrome, an important pigment whose role in plant development will be discussed in the next chapter. If chloroplasts do not receive appropriate illumination at an early stage of their development, they become abnormal in structure, forming a colorless network of tubes instead of green pigmented layers. Once they have developed in this manner, they cannot ever assume normal morphology; thus, for proplastids to become chloroplasts, they need light of the right wavelength delivered at the right stage of their development.

Like all cells and cell organelles, chloroplasts are surrounded by membranes that are differentially permeable—that is, the membranes permit some substances to pass through, while restraining others. All membranes are composed of two major structural components: fatty substances and proteins, which are large molecules formed by linking any of 20 component amino acids. For any substance to pass through a membrane, it must find either a pore big enough to slide through or a receptor protein embedded in the membrane whose outer surface can bind that specific substance. With the aid of stored chemical energy, the receptor protein can change shape and move the substance into the interior of the cell. In this way, membranes can control the composition of spaces they surround.

Inside the outer membrane of the chloroplast is a large space, the stroma, in which is found a second membrane-bounded space in the form of a network of interconnected green membranous discs, the thylakoids, with liquid-filled centers, or lumens. All of the photosynthetic pigments in the chloroplast are found in the thylakoids, some of which are organized into stacks called grana (singular *granum*) and others into unitary chains linking the grana. Together, these interconnected thylakoids form the photosynthetic layers of the chloroplast.

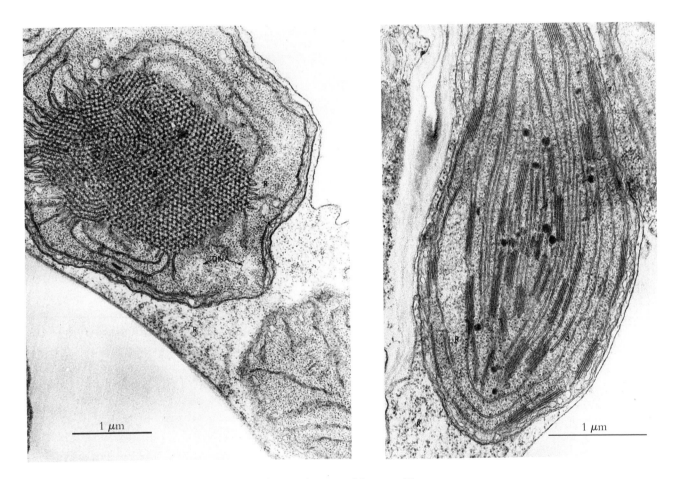

Left: A maize proplastid from a dark-grown plant is dominated by a netlike prolamellar body. Under the influence of light, this structure forms thylakoids and grana. *Right:* A maize chloroplast from a light-grown plant has formed parallel layers made of thylakoids and the stacks of thylakoids that make up grana.

The structure of the chloroplast can be considered to consist of enclosures within an enclosure. Inside the stromal region of the chloroplast lie the thylakoids, which are made of pigment-containing membranes surrounding a smaller space called the lumen.

CHLOROPLAST

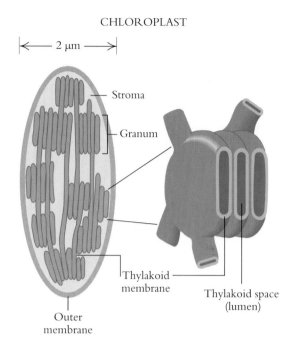

The membranes of thylakoids have characteristics that differ from those of the outer membrane in ways that are important in the process of photosynthesis. Some portions of the proteins in thylakoids consist of chains of hydrophobic (water-avoiding) amino acids that become located preferentially in the fatty parts of membranes, while the more hydrophilic (water-loving) parts of the amino acid chains extend into the interior of the thylakoid disc on one side and into the stroma of the chloroplast on the other. Throughout biology, proteins concerned with energy transduction have this type of transmembrane structure; in the thylakoid, such proteins can transfer captured light energy to the chemical reactions that eventually dismember a water molecule and create stored energy.

Light energy is converted to a chemical form only at specialized reaction centers, located in the stacked thylakoids of the grana. Only a small fraction of the pigment present in the chloroplast is found in such reaction centers; the remainder of the chlorophyll molecules, called antenna chlorophyll, can absorb light energy, but are unable to perform useful chemical work by themselves. Instead, the antenna pigments pass on absorbed light energy to a closely placed neighboring molecule of pigment through an efficient process called resonance energy transfer. This process is often compared to the activation of an untouched tuning fork by the vibrations of a nearby tuning fork. When the transmitted energy arrives at a chlorophyll molecule in the reaction center, it is able to perform chemical work, because the reaction center contains the necessary enzymes, structural proteins, and smaller molecules.

Through the capture and transmission of light energy, antenna chlorophyll increases the efficiency of the chloroplast. Even in bright sunlight, the average chlorophyll molecule absorbs only a few light quanta per second. This is not enough light energy to accommodate the rate at which carbon dioxide can be fixed in the dark reaction of photosynthesis. It would obviously be inefficient for a large and complex reaction center to sit partly idle because of an insufficient rate of energy flow. The resonance transfer of energy from abundant and widespread antenna pigments to the relatively few pigment molecules located at reaction centers keeps photosynthesis proceeding at a good pace.

While the chloroplast is in many ways a uniquely independent organelle in the cell, even possessing its own DNA, its development is tightly linked to the rest of the cell. For example, some nuclear genes can affect chlorophyll synthesis, and chloroplasts must import many of their impor-

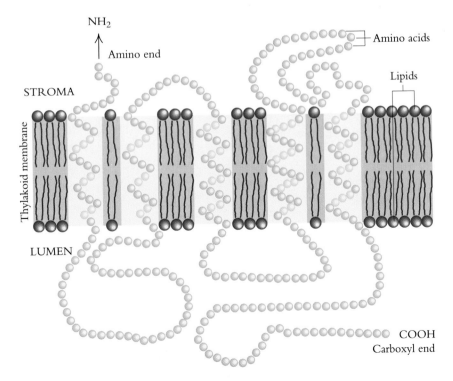

A thylakoid membrane comprises two layers of lipids with hydrophilic heads and fatty acid tails. The hydrophobic part of a transmembrane protein sits in the fatty layer, while the hydrophilic ends protrude into the lumen and stroma.

25 °C

17 °C

+/+   *hcf/hcf*   *hcf/0*   +/0

Left: Plants must contain appropriate pigments to absorb photosynthetically active light. The albino corn seedling (*left*) cannot photosynthesize; the normal seedling (*right*) can. *Right:* A wild type plant (+/+) has normal pigment content; a mutant with two copies of the mutant gene *(hcf/hcf)* has reduced pigment, especially at 17 °C; a mutant with one copy (*hcf*/0) has reduced pigment at both temperatures; a plant with no mutant genes but only one copy of the normal gene (+/0) has almost normal pigment content.

tant protein constituents from the cytoplasm. One very important chloroplast enzyme involved in carbon dioxide fixation, ribulose bisphosphate carboxylase (Rubisco), is an especially instructive example of cooperation between the chloroplast and the rest of the cell. Rubisco is made of two protein subunits, the larger encoded by chloroplast DNA and synthesized in the chloroplast, the smaller encoded by nuclear genes and synthesized in the cytoplasm. After synthesis, the smaller subunit is transported into the chloroplast, where it joins with the larger subunit to make the active enzyme. Thus, only through cooperation among chloroplast, nucleus, and cytoplasm can carbon dioxide fixation be accomplished.

Although chloroplasts cannot replicate independently of the rest of the cell, it is widely believed that they originated as free-living photosynthetic microorganisms, resembling the modern "blue-green algae," more properly called cyanobacteria, that form pond scums. This belief is based on the fact that chloroplast DNA resembles the DNA of bacteria, rather than that of the plant cell's nucleus. In addition, the chloroplast's machinery for synthesizing proteins closely resembles that of bacteria, rather than that of the nucleated cell. It is hypothesized that some primitive cyanobacterium-like cell came by chance to lie within a complex unpigmented cell, probably by an infolding of a portion of the larger cell's outer membrane against which the bacterium had come to lie. Once inside the host cell, the bacterium was able to continue its own replication by virtue of its own DNA, and it was able to absorb and metabolize materials from the host cell's cytoplasm and to secrete its own materials into the cytoplasm.

Eventually, a true symbiosis developed between the bacterial and host cells. The bacterium supplied photosynthesized nutrients to the cell, which in turn bathed the bacterium in a complex soup of its own varied organic and inorganic molecules. During many cycles of replication, the chloroplast could safely lose some of its independent functions by random mutation, if those functions were replaced by those of the host cell. In fact, it became advantageous for the bacterium to depend on the products of its host's nuclear genes and cellular metabolism. Why should the bacterium waste energy synthesizing its own molecules and cellular structures when they would only duplicate those provided free of charge by the host cell? Through a gradual loss of function and a growing dependence on the host cell, the structure and chemistry of the original cyanobacterium changed progressively. After its reproductive cycle became synchronized with that of its host cell, it became a chloroplast, a stable organelle in the cell.

The Light Reactions

Some years after the discovery of separate light and dark reactions, Emerson and Charlton Lewis of Caltech were able to show that even the rapid light reaction is complex and must consist of at least two component processes. When they gradually increased the intensity of red light of about 680 nm on a leaf, the rate of photosynthesis increased as well until it came to a saturation point, beyond which increasing the intensity had no

A cell containing four chloroplasts, each with many included starch grains. The large open space is the vacuole; the finely granular layer is the cytoplasm.

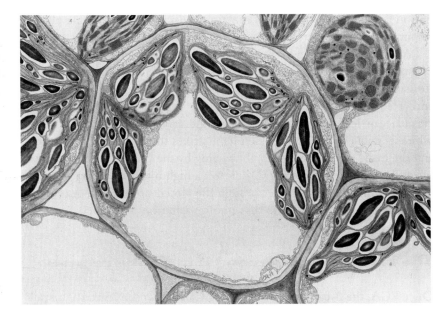

itself is immobile, but it can be broken down into many molecules of simpler, transportable sugars through the action of certain plant and animal enzymes. Food and energy can be stored as starch until needed, when the starch is broken down and the simple sugars formed are transported to the location where they are required. The polysaccharide cellulose, by contrast, is a tough and durable structural component of plant cell walls. Cellulose cannot be degraded by the usual plant or animal enzymes, and is therefore unavailable to plants and animals as a reserve foodstuff; it is, however, readily broken down to simple sugars and oxidized by certain fungi and bacteria. Such microorganisms help ruminants and termites to digest and consume cellulose. As we shall see in Chapter 3, some plant cells do form cellulose-digesting enzymes just before leaf fall.

Hexose may be converted either to sucrose or to starch grains, depending on the genetics of the plant and on specific regulatory processes within the leaf. Whereas sucrose is formed in the cytoplasm, starch grains are formed only within plastids, a group of related organelles that includes chloroplasts. There may be one or several grains within a plastid, and plastids may become so choked with enlarging starch grains that their structure is disrupted. Special starch-containing plastids called amyloplasts are abundant in storage organs such as the potato tuber; these plastids do not contain photosynthetic thylakoid layers but are instead colorless. The biochemical machinery leading to the synthesis of starch in plastids is similar, however, whether or not the plastids are photosynthetic.

The giant, insoluble starch molecule serves as an effective storehouse for the photosynthesized sugar because, sequestered in a plastid, it is effectively removed from the metabolic flux and flow of the cytoplasm. Were the equivalent amount of sugar to remain dissolved in the cytoplasm, its concentration would be huge, as in maple syrup. Such a buildup of sugar inside the cell would cause a massive movement of water into that cell, since water molecules diffuse readily through membranes, but sugar molecules do not. Once water molecules are inside the cell, their passage out again is hindered by the interference of so many sugar molecules. Thus, if pure water were outside the cell and a concentrated sugar solution inside, many more water molecules would penetrate each unit of membrane from the outside than from the reverse direction. Over time, pressure would build up as the volume of water inside the cell increased and the cell contents were pressed against the rigid wall.

The diffusion of water molecules through a membrane in response to concentration differences on the two sides of the membrane is called osmosis. It accounts for the uptake of most of the water absorbed by plant cells as well as the movement of water from cell to cell within the plant. Osmotic pressure keeps leaves and other plant organs rigid. When loss of water exceeds uptake, the pressure of the cell contents on the wall decreases, and cells collapse like deflated balloons. The plant wilts.

Stored starch is used as an energy source by the plant when it is unable to photosynthesize. For example, seeds must germinate and grow before producing photosynthetically active leaves; to get started, they often rely on stored starch. Starch is also found in storage organs such as tubers and

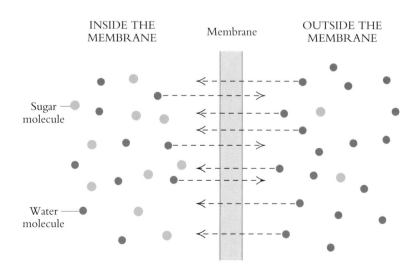

INSIDE THE MEMBRANE

Membrane

OUTSIDE THE MEMBRANE

Sugar molecule

Water molecule

Osmosis is the diffusion of water molecules through a differentially permeable membrane. In this example, sugar molecules inside the vacuole dilute the concentration of water molecules and restrain their passage through the membrane. The net movement of water is thus to the inside.

rhizomes, where it furnishes an energy source for the production of new vegetative organs after an overwintering or other dormant season.

The decision to make sucrose or starch appears to be regulated by small phosphorus-containing molecules such as free phosphate, triose phosphate, and the hexose sugar, fructose 2,6-bisphosphate. Their concentrations are in turn regulated by the activities of enzymes called kinases, which add phosphate groups to molecules, and other enzymes, called phosphatases, which remove phosphate groups from molecules. The activity of these enzymes is regulated by genes and certain small molecules, including phosphate groups themselves.

The Control of Photosynthetic Activity

For a plant to carry out vigorous photosynthetic activity, it must obviously have available abundant light energy, an adequate supply of carbon dioxide, and plenty of water. Under the best conditions in the field, leaves are adequately supplied with all three requirements. In less than ideal conditions, however, the process is said to be limited by that factor which is in relatively short supply.

When a low light intensity limits the rate of photosynthesis, steadily increasing the light intensity leads to a steady increase in the photosynthetic rate until the system becomes light saturated. Adding light energy after this point can no longer increase photosynthetic yield, and may actually harm the photosynthetic apparatus by generating damaging reactive chemicals. The light intensity at saturation depends on the previous history of the plant: plants grown under high light intensities will have formed a greater number of photosynthesizing cell layers, more chloroplasts per cell, and more chlorophyll per chloroplast. Such plants will require correspondingly more light to reach saturation.

At light saturation, the rate of photosynthesis can be further increased by supplying higher levels of carbon dioxide. This effect, too, will reach saturation. When neither more carbon dioxide nor additional light energy will further increase photosynthetic yield, the leaf is probably limited by internal factors rather than by the environment. It cannot carry out photosynthesis faster than the level of chlorophyll, the number of chloroplasts per cell, and the number of layers of photosynthetic cells in the leaf will allow. The ability to mobilize sugars into stored starch or into sucrose for transport out of the leaf may also limit the rate of photosynthesis, because the buildup of sugar inhibits the rate at which new sugar can be formed. The plant has efficient mechanisms for adjusting each of these internal parameters to meet changing needs.

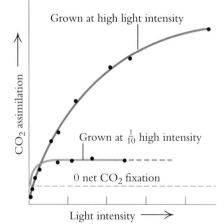

An increase in light intensity causes a greater increase in the photosynthetic rate of plants previously grown at higher light intensity.

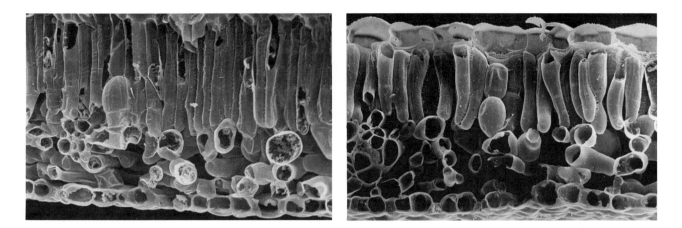

Scanning electron micrographic views of cross sections of sun *(left)* and shade *(right)* leaves of the same tree, a maple. Note the greater development of palisade cells in sun leaves.

The synthesis of chlorophyll may be limited, for example, by the availability of nitrogen or magnesium, which are structural components of the chlorophyll molecule. If these materials are added to the soil or if roots grow into new areas of soil where these elements are abundant, the plant will be able to synthesize more chlorophyll molecules and thus create more photosynthetic reaction centers. The number of chloroplasts per cell is less easily altered, but under some conditions, including high light intensity, the cell can produce additional chloroplasts. Plants grown in bright sunlight will tend to have thicker leaves than plants grown in shade, because sunlight stimulates the addition of extra layers of palisade cells containing many chloroplasts.

Even the orientation of chloroplasts in the cell is regulated—especially by light. Recall that the chloroplast is a wafer-shaped body with a relatively broad surface area, but a thin edge. If the thin edge is presented to the incoming light, photon capture is minimized, but if the chloroplast is rotated so that its maximum surface area is exposed to light, the absorption of light will increase. Chloroplasts tend to present their thin edge to incoming light of high intensity, and their broad surfaces to light of low intensity. Chloroplasts also adjust to changes in light by redistributing themselves in the cell. In dim light, chloroplasts will spread over the cell surface, whereas in bright light they will stack up along the cell's inner edge, in effect shading one another from sunlight.

Chloroplasts change their orientation by being pulled and tugged along cytoplasmic strands called microfilaments. These strands contain the

Chloroplasts in the cells of duckweed distribute themselves to optimize light absorption in dim light and minimize light absorption in strong light.

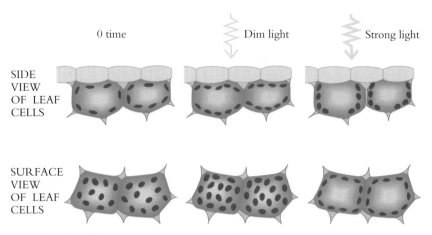

protein actin, also present in muscles. This protein contracts in the presence of calcium to provide the motive force for moving chloroplasts. We will discuss this further in Chapter 4.

As the image of the sun moves across the sky during the day, the entire leaf can change its orientation to maximize light absorption. Sun tracking is well known in sunflower stems, but the leaves of many species can adjust their orientation by means of fleshy organs called pulvini (singular *pulvinus*) found at the base of the petiole. The leaf is made to move relative to the stem by the osmotic swelling and shrinking of cells on opposite sides of the petiole near its base. If, for example, the upper cells swell and the lower cells shrink, the leaf will move downward, while a reversal of these swellings will raise the leaf. This mechanism can also move the leaf from side to side, ensuring flexibility of leaf orientation.

The final arbiters of the leaf's overall photosynthetic activity are undoubtedly the stomata. These pores are generally present in abundance on both the upper and lower epidermis of the leaf. Each stoma is surrounded by a pair of guard cells whose turgor alternates with sun and shade or hydration and dehydration. When the guard cells become turgid they swell, opening the stomatal pore and permitting the free diffusion of gases between the cavity below the stoma and the outer environment. The pore opens because the outer wall of the guard cell is thinner than the inner wall; under pressure, the outer wall bulges, much as a weak spot on a balloon bulges outward when the balloon is inflated. This outward bulging carries the thicker inner wall of the cell along with it and the pore is opened. When guard cells lose their turgor, the thick inner wall contracts, pulling the thin outer wall with it. The stomatal pore closes, effectively sealing the leaf against gas exchange. When all the stomata are closed, photosynthesis is limited to the CO_2 produced by the leaf itself as it respires.

Except for crassulacean plants, stomata are open in the light and closed in the dark, probably because the guard cells themselves are capable of carrying on a rapid photosynthetic fixation of CO_2. By this means, the cells generate ATP, which is used to energize ion pumps located in the cell membrane. These pumps are proteins able to facilitate the passage of potassium and chloride ions into the guard cell from adjacent subsidiary cells against their concentration gradient. The accumulation of these ions causes an osmotic influx of water, thus leading to greater turgor and stomatal opening. In darkness, when the energy supply is diminished, the reverse processes occur spontaneously, closing the stomata.

In addition to being driven by photosynthetic light absorption, stomatal ion pumping also seems to be stimulated by a particular reaction driven by blue light. The pigment receiving this blue light energy is probably a flavoprotein (a protein containing riboflavin, also known as vitamin B_2) found in the membranes of the guard cells. The exact composition of this pigment is obscure, and its mode of action is even more mysterious. Most biologists hypothesize that the flavoprotein is one component of the cell's chain of electron carriers, and that its activation by light somehow causes more protons to be sent through the ATP-synthesizing site of the guard cell. At the moment this theory is pure conjecture in need of experimental verification.

Photosynthesis is a remarkable, highly regulated process for harnessing the energy of the sun's photons. The complex architecture of the plant and the incredibly intricate biochemical and genetic controls that regulate photosynthetic activity may be viewed as refinements of the basic process of trapping the photon and converting its energy into chemical form. Without these refinements the process would be inefficient, but with them the leaf is able to capture and use up to 5 percent of all the energy falling on its surface—enough to support the vigorous plant growth that feeds all animal life on earth.

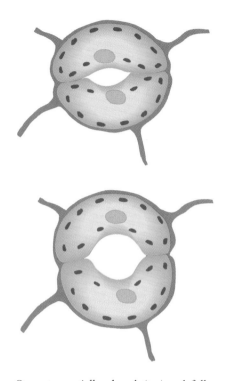

Stomata partially closed *(top)* and fully open *(bottom)*, a consequence of the internal turgor of the guard cells.

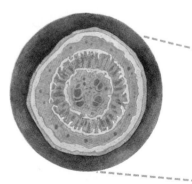

For plants growing in the earth's temperate zones, timing is the essence of survival. The rooted green plant cannot move to a better location when a change in the season causes the temperature, light, availability of water, or other environmental condition to alter for the worse. The plant must synchronize its vital activities to occur when climatic conditions are favorable, generally in the spring or summer. In the winter or arid summer, it may lapse into relative inactivity or even complete dormancy.

Most annual plants adjust to the change in seasons by alternating an active vegetative stage, in which the plant consists exclusively of roots, stems, and leaves, with a temporarily dormant stage, usually a seed formed after the plant has flowered. By contrast, most biennial or perennial plants survive the winter by shedding succulent and tender parts that might be injured through freezing or desiccation, then modifying the remaining parts to withstand the season's adverse conditions. Both strategies require plants to anticipate the future, so that they may initiate protective measures well in advance of hard times to come and to reinitiate growth when the danger has passed. Their secret is to estimate the date by measuring the length of day, then to translate that information into chemical or physical growth controls calibrated to last as long as they are needed.

TWO

Light and the Life Cycle

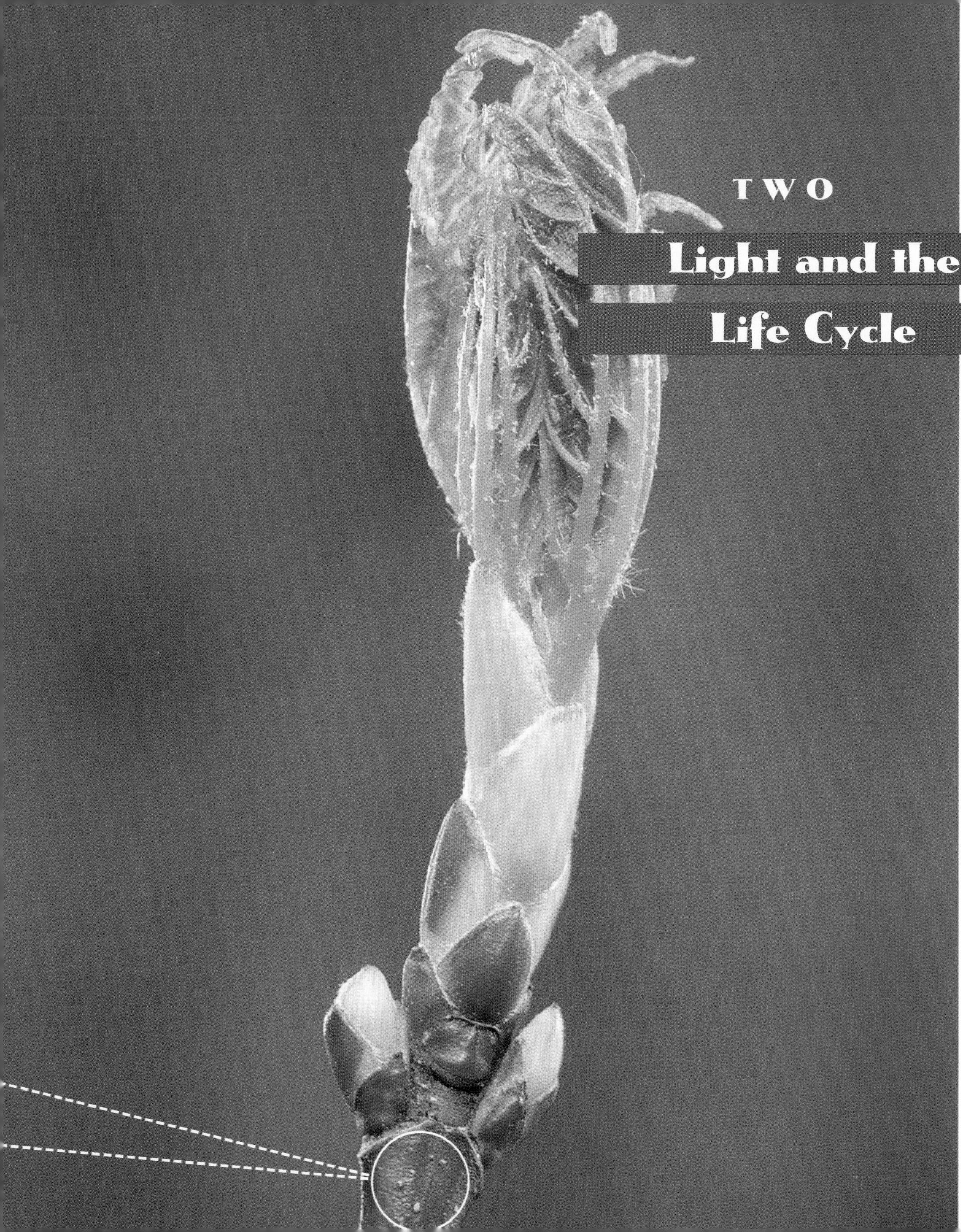

Strategies for Overwintering

Annual plants germinate in the spring, flower during the same year, then die, either because they have followed a plan of programmed senescence and death or because they have fallen victim to the sudden winter cold. Although the annual plant body has perished, its progeny survive most harsh winters in the form of dormant seeds. In contrast, a biennial plant lives through one winter before producing seeds. In the first year, a typical biennial plant produces a whorl of leaves close to the ground, called a rosette, and an underground storage organ; in this form, the plant safely overwinters, usually blanketed by snow. In its second year, it adopts a strategy similar to that of the annual plant: it flowers and dies before the cold weather, leaving dormant seeds behind to overwinter.

Perennial plants survive the months of freezing temperatures by donning winter armor. The well-protected winter buds of many trees, shrubs, and herbaceous perennials, which insulate the delicate growing point from cold, must form before the first frost. The buds are usually surrounded by closely overlapping scale leaves that tend to conserve the interior heat; they are sealed against the outside world by a combination of waxy and shellaclike materials that prevent the entry of freezable water. In addition, winter buds are less succulent than other plant parts. Less free water is present than in growing buds, and winter buds contain specific protein and other molecules that "bind" water, making it considerably less likely to freeze. (Water that hydrates flour into dough is a familiar example of bound, partly unfreezable water.) Some winter buds and other plant parts even synthesize special "antifreeze" chemicals like the ones we put into our automobile radiators in the winter; these protect against ice formation because of their low freezing point. Clearly, the plant uses defensive tricks that make good sense to engineers, chemists, and designers!

No matter during which years in its life cycle a plant flowers, it blooms and produces seeds before the arrival of winter. Plants are genetically programmed to produce flowers at just the right time of year, at a particular latitude and altitude, to guarantee the highest number of flowers and seeds for reproduction. Clearly, it is best for the plant, before it flowers, to grow to a size capable of supporting the greatest number of fruits and seeds feasible. If it flowers too soon, then the vegetative body will be too small to support many flowers, and if it flowers too late, the fruit and seed may not ripen before the first killing frost. Each wild plant has evolved its own best reproductive schedule, and all cultivated plants must be bred to flower at an appropriate time in the region of their use.

Many plants flower and produce seeds in the late summer or autumn. If seeds produced late in the year were to germinate immediately, the tender seedlings would die at the first severe frost, or at least experience considerable injury. To postpone germination until spring, plants of the temperate zone have evolved numerous mechanisms to ensure that when the seed matures, it is unable to start growth immediately. The seed is prevented from germinating by built-in restraints that are timed to disappear when conditions become more favorable.

Timing and the Germination of Seeds

The seeds of many plants are physically restrained from germinating by the seed coat, the outer covering that envelops and protects the embryo and its supply of food stored in the endosperm. The seed coat is made up of several layers of sturdy, thick-walled cells, and in many plants is impregnated with wax or a varnishlike material. Such a seed coat might, for example, be impermeable to water, without which seed germination is impossible. Thus, even though the seed is placed in moist, warm conditions, favorable for germination and growth, it will remain dormant until the integrity of the impermeable seed coat is breached. Microbes may slowly degrade the seed coat while the seed rests in the soil for several months, or some mechanical event, such as physical abrasion, may rupture the coat.

In arid zones, the seed coat is often made to rupture just when water is plentiful. A seed subjected to a light rainfall, which would evaporate too

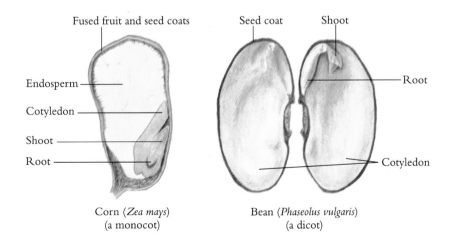

Corn (*Zea mays*)
(a monocot)

Bean (*Phaseolus vulgaris*)
(a dicot)

All seeds are survival packets, containing an inactive embryonic plant, a source of food (the endosperm), and a protective coat. Monocots, such as corn, and dicots, such as beans, differ in basic structure. In the bean, the stored food is in the fleshy cotyledons.

quickly to ensure survival, remains undisturbed and dormant in the soil. However, in a more substantial downpour, the seed is tumbled about in temporary rivulets and rubbed against sharp-edged sand or gravel particles that abrade or puncture its coat. Water enters and the seed germinates. A root forms quickly, penetrating the lower soil horizons and tapping their water reservoirs before the upper soil dries out completely.

Germination in such seeds can be promoted artificially by any mechanical force strong enough to rupture the seed coat. In the process of scarification, the seed coats of commercial crops are deliberately punctured by needles or sandpaper before planting to ensure that the seeds germinate all at once. To break down the barrier to water entry, some especially hard and impermeable seeds are even exposed briefly to vigorous chemical agents like concentrated sulfuric acid! Unless treated in some such drastic fashion, these seeds can remain dormant for a very long time. Dormant seeds of the lotus (*Nelumbium nelumbo*), buried for about three centuries in the dry loess soil of China, have been successfully germinated after scarification. The longevity of such seeds seems to be limited only by the ultimate depletion of their stored food reserves. Such reserves are slowly used up in respiration to obtain energy for the maintenance of the seed's cellular structure.

Sometimes, the barrier to germination arises not from the impermeability of the seed coat to water, but from the impermeability to oxygen of the testa, a thin layer of tissue inside the seed coat. For example, the fruit of *Xanthium strumarium,* a common annual weed called cocklebur, contains two seeds, both structurally mature and apparently ready to germinate. Yet only the upper seed of this pair will germinate in the first year, while the second waits until the following year or even later. However, if the fruit is placed in air with an elevated oxygen content, then both seeds are likely to germinate. It appears that the upper seed testa is more permeable to oxygen than is the lower; thus it can germinate promptly in atmospheric oxygen, while the lower seed cannot. Raising the oxygen concentration seems to overcome the higher oxygen permeability barrier of the lower seed testa. The second seed will also germinate if the testa is punctured with a needle or artificially removed from the seed. Alternatively, if the fruit is left in the ground for a season, soil microorganisms break down the testa of the second seed, permitting it to germinate in ordinary atmospheric oxygen. The species's chances of survival in the long run are enhanced by staggering the production of its offspring in time, rather than risking all progeny at once. To ensure a staggered schedule of seed germination, the plant can also use chemical tricks to extend dormancy, as we shall see in the next chapter.

Cocklebur fruits contain two seeds encased in a tough fruit coat armed with hard, hooked barbs. By sticking to animals, the barbs facilitate the seeds' distribution.

In many plants, the germination of seeds is controlled by light. For example, it has long been known that some seeds will not germinate when left in total darkness. Such *photoblastic* seeds will begin germinating after brief exposure to white or, especially, red light. *Photodormant* seeds behave in the opposite way; they will germinate in darkness but are inhibited by light. In both instances, the pigment in the seed that absorbs the effective reddish light is *phytochrome,* a blue-green protein that probably exists in minute quantities in all plant cells. Phytochrome controls an amazing number of physiological processes throughout the life cycle of the plant, from flowering and dormancy to growth and even senescence. It sometimes acts through growth hormones like gibberellin or abscisic acid, which can promote and restrain, respectively, the germination of certain seeds, even without light.

How could the photocontrol of seed germination by light be useful to the plant? If a seed lies on the surface of the soil, it is in danger of desiccation and mechanical injury once it germinates. If the seed is photodormant, then it will not germinate until it has been placed in the dark through burial under a protective layer of soil. Conversely, a photoblastic seed germinating too deeply in the soil might run out of food reserves before reaching the surface, where it can absorb light and begin photosynthesis. Since the porous surface soil permits some light penetration, the photoblastic seed can be light-activated through phytochrome to germinate when it is at a suitable distance below the soil surface. This mechanism works well because only minute quantities of light energy suffice to set off germination. Somehow, the results of phytochrome activation are greatly amplified within the plant; we shall examine how this might be accomplished at the end of this chapter.

Flowering and the Discovery of Photoperiodism

Horticulturists have long noted that no matter when the seeds of a species are planted, flowers appear at about the same date. Dividing cells at the apex of the stem cease creating vegetative, leaf-forming buds and instead form the reproductive buds from which flowers develop. The regularity in the onset of flowering tells us that flowering is not dependent solely on the passage of time, on the mere attainment of some critical size, or on other developmental events such as the storage of sufficient starch. Some early French and German plant physiologists suspected that seasonal variations in light might be the factor controlling the onset of the reproductive state, but it was not until about 1920 that the role of daylength in the control of flowering was firmly established.

Wightman Garner and Henry Allard, working at a tobacco experiment station in Maryland run by the U.S. Department of Agriculture,
were interested in maximizing the harvest of this plant. Tobacco is cultivated for its leaves, and these investigators were delighted to find that
some extraordinarily large plants with giant leaves had appeared as if by
accident in their experimental plot. They named this aberrant form Maryland Mammoth and waited for it to flower, so that they might collect its
seeds for further experimentation. To their distress, Maryland Mammoth
remained vegetative long after the smaller plants in the field had flowered
and produced seed. Fearing that the approaching first frost might harm
this potentially valuable plant material, they dug up their Maryland Mammoth plants and transported them to a greenhouse. There the plants continued to grow vegetatively until almost Christmas, when fortunately they
flowered and set seed.

When the progeny repeated the behavior of the original plant, failing
to flower in the field during the normal growing season but flowering in
the greenhouse in mid-December, Garner and Allard set about perform-
ing systematic experiments that would explain such behavior. They suspected low light intensity, the light quality, the interplay of temperature
and light, and other factors, but finally became convinced that the controlling stimulus was the short daylength of December. A simple experiment confirmed their hypothesis: plants on flatcars were moved on rails
into either an illuminated greenhouse or a darkened garage to simulate

Garner and Allard's short-day Maryland Mammoth tobacco in flower
(*left*) and in the vegetative state (*right*).
Flowering during the short winter
days occurred naturally, and could be
prevented by extending the daylength
with artificial illumination.

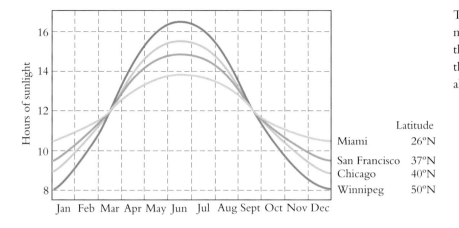

	Latitude
Miami	26°N
San Francisco	37°N
Chicago	40°N
Winnipeg	50°N

The change from maximum to minimum photoperiod is greater the farther one gets from the equator. At the equator, the daily photoperiod is always 12 hours.

days and nights of various lengths. At Maryland's latitude, about 9 hours of light are combined with 15 hours of darkness in late December. Only when the plant-carrying flatcar stayed in darkness and light for periods mimicking late December did the plants flower.

Garner and Allard later found that certain plants, like tobacco and soybeans, flower consistently after exposure to a certain number of consecutive short days, while others, like spinach and most cereals, respond to long days, and still others, like the tomato, form flowers spontaneously, without reference to the length of day at all. They named this phenomenon, the response of plants to the relative lengths of day and night, *photoperiodism,* and classified plants as short-day, long-day, and day-neutral, according to their flowering behavior. The tomato and other day-neutral plants flower in response to a genetic program that directs them to produce a certain number of leaves before producing flowers.

The hours of light in a day define the photoperiod. It changes with season because the axis of the earth's rotation is not perpendicular to the plane of its annual revolution around the sun, but is instead inclined at an angle of about $23\frac{1}{2}°$. This inclination makes no difference at the equator, where there are always 12 hours of light and 12 hours of darkness, but it makes a progressively greater difference as distance from the equator increases. In the temperate mid-latitudes, daylength varies from about 9 to 15 hours over a six-month period, while at the poles seasonal periods of continuous light or continuous darkness last for many days. When daylength at any latitude is plotted against date, the resulting curve resembles a sine wave whose amplitude depends on the latitude. The closer one comes to the equator, the less is the seasonal change in the length of the photoperiod. Twice each year, at the vernal and autumnal equinoxes, the orientation of the earth toward the sun is such that the light and dark halves are divided exactly by the earth's axis of rotation. On these days, the photoperiod is 12 hours all over the earth. At the summer and winter solstices, daylength reaches its maximum and minimum, respectively. The daily change in photoperiod is not constant over time; it augments or diminishes very slowly near the minimum and maximum photoperiods, and very much more rapidly in between.

Plants have evolved to compete effectively with other plants at the same latitude. For example, some species of range grasses that can grow at latitudes from Texas to the Canadian border have formed a large number of ecological variants adapted to the photoperiods within a narrow latitude. Both the daylength at which these variants begin to flower and the daylength at which flowering and seed yield are greatest vary with their geographical locations. Thus, the botanist seeking to transplant, say, a Texas variant to Iowa will run into trouble, because flowering may begin on a date not well suited to the new location. For this reason, botanists take photoperiodic behavior into account when trying to introduce a new plant into the agriculture of a particular region. The same reasoning explains why seed companies offer so many different varieties of the same crop to farmers. Even a small increase in yield because of a better match with local conditions can mean the difference between a successful harvest and crop failure.

Is There a Flowering Hormone?

The photoperiodic behavior of plants has been intensely analyzed over the last half-century, and we now understand many of the mechanisms involved. Plants perceive the photoperiodic signal with a precision that is astounding. Some short-day and long-day plants can perceive a difference of from 2 to 5 minutes in the length of day. For example, the common cocklebur is a short-day plant that will flower if the photoperiod is less than $15\frac{1}{2}$ hours per day, but not if it exceeds that value by more than a few minutes. We know that the plant measures the dark part of the daylength cycle, for if a cocklebur is placed in an artificially illuminated chamber where dark and light periods can be varied separately, then an uninterrupted $8\frac{1}{2}$-hour dark period turns out to be essential for flowering, but the continuous $15\frac{1}{2}$-hour light period does not. If the $8\frac{1}{2}$-hour dark period is interrupted by even a momentary flash of light or by a prolonged exposure to very dim light, the plant will not flower. Even bright moonlight or

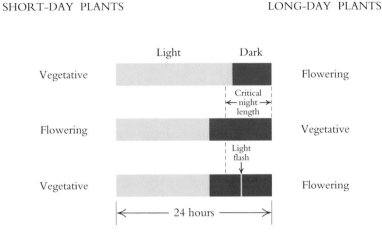

BEHAVIOR OF
SHORT-DAY PLANTS

LIGHT REGIME

BEHAVIOR OF
LONG-DAY PLANTS

Light   Dark

Vegetative   Flowering

Critical
night
length

Flowering   Vegetative

Light
flash

Vegetative   Flowering

← 24 hours →

The critical night length determines flowering behavior in both short- and long-day plants. Thus, they respond in opposite ways to a light flash in the middle of the dark period.

the striking of a match can interrupt this delicate process. In long-day plants, the length of the dark period also determines the onset of flowering, but light interruptions stimulate flowering rather than suppress it.

The length of day that just demarcates the boundary between the vegetative and flowering states is called the critical photoperiod. Short-day plants flower when the photoperiod is at or below the critical length, and long-day plants when the photoperiod is at or above the critical length. It is this characteristic, not the absolute length of day, that determines the photoperiodic type. Thus, the cocklebur, a short-day plant, has a critical photoperiod that is several hours longer than that of black henbane (*Hyoscyamus niger*), a long-day plant. The short-day cocklebur reaches its critical photoperiod during the decreasing days just after midsummer, whereas the long-day black henbane reaches its critical photoperiod during the lengthening days in early spring.

Most plants need only brief exposure to the correct photoperiod for flowering to be initiated and then maintained indefinitely. The cocklebur and the Japanese morning glory (*Pharbitis nil*) require only a single short-day cycle for flowering to begin, while most varieties of soybean require about four such consecutive cycles. Once the plant has received the appropriate stimulus, it continues flowering even if returned to inappropriate photoperiods. Such plants are said to be photoperiodically *induced*. In some plants, flowering is halted if the exposure to noninductive photoperiods is prolonged, while other plants cannot revert to the vegetative state once flowering has begun.

Although it is the bud that is transformed following photoperiodic induction, it is the vigorously growing, nearly mature leaf that receives the photoperiodic stimulus. Using an opaque barrier to separate one part of the plant from the rest, researchers are able to expose only part of the plant to an inductive photoperiod. As long as mature cocklebur leaves are exposed to inductive photoperiods, it does not matter that the bud is not. Thus, for flowering to be induced, some stimulus must move between the leaf and the bud; either the induced leaf exports a floral promoter to the otherwise vegetative bud or the uninduced leaf exports a substance that inhibits the bud from a "natural" tendency to flower. More than 50 years ago, the Soviet plant physiologist Mikhail Chailakhian proposed that a specific flower-inducing hormone, which he named florigen, is exported from induced leaves to receptor buds.

The most convincing experiments favoring the existence of florigen have been grafts. When grafts are made, plants are cut and joined in such a way as to establish direct connections between the vascular tissues of the two graft members. When an induced cocklebur is grafted to a vegetative cocklebur, both plants will flower. It is even possible to induce the flowering of a vegetative receptor by grafting on a single induced leaf.

A successful graft will establish continuity between the living cells of the phloem tissues of donor and receptor. The floral stimulus probably travels out of a leaf and into a bud along with the mass flow of nutrients, mainly sugars, through the phloem. Since this tissue is known to transport various organic solutes, the evidence is consistent with the existence of an organic floral promoter in the phloem sap. Yet, almost all attempts to extract specific flower-inducing substances have failed, and even the "successful" attempts have been difficult to repeat. Nonetheless, reports of flower-inducing extracts continue to appear, and it is possible that one of these reports will correctly identify florigen. To qualify for that designation, the substance must be produced only in induced plants (or at least in greater quantities in those plants). In addition, because the floral stimulus has been shown to be transmitted between a short-day tobacco and a long-day henbane, the substance must be capable of inducing both short-day and long-day plants.

The floral stimulus moves more readily through the phloem if bright, photosynthetically active light is administered to a donor leaf after the leaf has been exposed to an inductive dark period. Transport through the phloem depends on the osmotic pressure built up in special cells stacked

The long-day plant black henbane (*Hyoscyamus niger*) in flower. This plant can send a flowering signal through grafts to short-day Maryland Mammoth tobacco, and vice versa.

to form pipelike structures, called sieve tubes, which connect the leaf with other parts of the plant. When the sugars produced by photosynthesis in the induced leaf are loaded into a sieve tube, the entry of water by osmosis builds pressure in the cell that squeezes the contents of the tube. The unloading of sugars at the tube's receiving end, such as a bud, diminishes pressure at that end, creating a pressure gradient from "source to sink" that drives transport from leaf to bud. The floral stimulus is carried along passively in this wave of transport. Each leaf competes with all other leaves in exporting its products to distant parts of the plant. Thus, if a photosynthesizing, noninduced leaf lies between the induced donor leaf and a receptor bud, then the transport of the flowering stimulus from donor to receptor may be inhibited. In such cases, the noninduced leaf feeds into sieve tubes that do not contain the floral stimulus.

Because no flowering hormone has ever been isolated, many researchers now prefer other explanations of flower induction. They propose that flowering is initiated through the interaction of several known substances that do not regulate a single process but work by affecting general growth processes. These known substances, to be discussed in the next chapter, include plant hormones gibberellin and ethylene, which can induce flowering in some but not all plants.

The First Spectral Analysis

After identifying the phenomenon of photoperiodism, the next step was to gain some understanding of the internal events governing photoperiodic response. To that end, in the 1940s a team of investigators working for the U.S. Department of Agriculture in Beltsville, Maryland, began a study of the effectiveness of different wavelengths of light in inducing flowering. This team, led by Sterling Hendricks, a physical chemist, and Harry Borthwick, a botanist who had worked on lettuce seed germination and dormancy, reasoned that the light active in regulating flowering must first be absorbed by a pigment, which might be found by matching its absorption spectrum with the action spectrum for the process.

To construct an action spectrum, one allows a particular wavelength band to fall on the organism, where it produces a measurable effect, such as the promotion of growth or the inhibition of flowering. One test at a selected wavelength is not enough, however, for the experimenter must account for the effects not just of the wavelength but also of the amount of energy received, which depends on the intensity of the light and the duration of the exposure. Obviously, a less effective wavelength at a high

Bünning's observations of the daily leaf movements in 10 varieties of soybean suggested that there was a close connection between circadian rhythms and the photoperiodic control of flowering. Those varieties that had pronounced rhythmic movements were strongly photoperiodic short-day plants, while those without pronounced movements tended to be day-neutral. Bünning proposed that a linkage exists between a light signal and an endogenous, circadian rhythm. According to his theory, phytochrome absorbs light that "resets the clock" by accelerating or de-

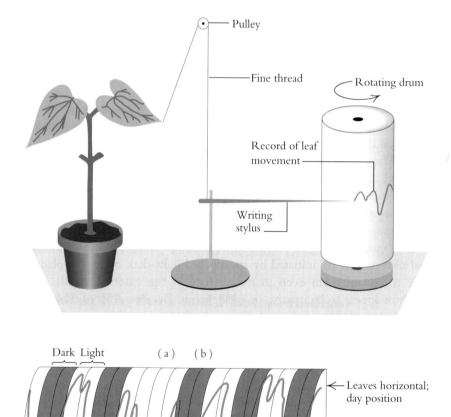

The circadian rhythmic movements of bean leaves can be automatically recorded on paper or film. Although the main change in the blade's orientation is from horizontal in light to vertical in darkness, other, smaller rhythmic movements are also seen. A 24-hour light period at (a) caused a resetting of the rhythm at (b).

laying the onset of the rhythm. Thus, the role of phytochrome would be to constantly adjust the endogenous circadian rhythm to keep up with seasonally changing daylengths. Most physiologists agree with Bünning, while noting that not all phytochrome-controlled events are linked to rhythms; etiolation and seed germination are two good examples.

Adjusting the circadian rhythm to the seasons keeps daily leaf movements on schedule, but how does the plant tell when it is time to prepare itself for winter, for example? Certain genes, the products of a long process of evolutionary change at a particular latitude and altitude, tell the plant when in its life history to expect a killing frost, and phytochrome, in conjunction with endogenous rhythms, tells the plant approximately when that date has arrived. The interaction of these two pieces of information, through largely unexplored mechanisms, tells the plant to prepare for the winter. Enzymes are synthesized for the production of special protective substances, and the shedding of leaves and the modification of growth patterns are initiated through changing levels of growth hormones and other regulatory substances that we will discuss in the next chapter. The result is a stripped-down, partially dormant, cold-resistant plant that can withstand the rigors of most winters.

In some biennial plants, the photoperiodic stimulus cannot be perceived until after the plant has been exposed to a period of low temperature. Because these plants cannot flower until after they pass through the winter, they avoid flowering in the late fall, when they would have no opportunity to complete their life cycle. After being exposed to winter cold, they undergo some sort of metabolic transformation that permits them to receive and respond to a photoperiodic stimulus, usually long-day, during the subsequent summer. Such plants are said to be *vernalized*. The stems of vernalized plants elongate greatly, or bolt, in response to photoperiodic induction. Bolting is usually, but not always, followed by the appearance of flowers.

Plants can be vernalized artificially by exposing their slightly hydrated seeds to temperatures of about 2 to 5 °C for at least four weeks in a refrigerator. Such vernalized seeds can be safely sown and will respond to photoperiod in the normal way. Cereal crops such as wheat or rye frequently exist in two varieties, designated spring and winter. Spring wheat, which requires no overwintering temperature before flowering, is sown where the growing season is long enough to yield a harvest in one year; it is planted in the spring and harvested in the fall. Where the growing season is too short for such a schedule, farmers plant winter wheat, which requires low temperature before it can respond to the critical photoperiod. Winter wheat is planted in the fall, overwinters as a low stubble,

The black henbane cannot bolt and flower in response to a long photo-period until it has experienced low temperatures, either in the field or in the laboratory. In nature, the plant is thus a biennial.

Long days   Short days

Not subjected to cold

Subjected to cold

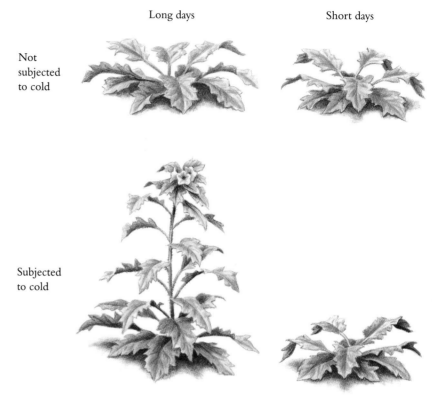

when it receives the required low-temperature stimulus, then flowers and produces grain the next summer, after having received the appropriate photoperiod in the spring. Grain can thus be grown over extensive areas of the globe by selection of the appropriate type. In some areas of the world, however, a short growing season compels the use of winter wheat, yet severe winters kill the plants while they are still stubble. In such regions, seeds vernalized in the laboratory may be planted in the spring, and a crop safely harvested that same year.

Amplifying the Phytochrome Signal

There is so little phytochrome in the average plant cell that no significant change in mass chemistry could result from transforming P_r to P_{fr}. Yet, a massive change in cellular biochemistry must result from the formation of P_{fr} in order for seeds to germinate, etiolated plants to become green, or flowers to form. Somehow the original phytochrome signal must be amplified. Logically, there seem to be three possibilities for amplification mechanisms: (1) phytochrome may activate enzymes, (2) it may alter membrane permeability, or (3) it may control the activity of genes. Each of these mechanisms has received some experimental support.

Phytochrome as enzyme

The transformation from P_r to P_{fr} might induce the inactive form of an enzyme, called a proenzyme, to convert to an active form. The original signal would then be amplified thousands of times as the enzyme catalyzed the same reaction over and over. It has been suggested that P_{fr} itself may act as an enzyme that attaches phosphate groups to particular other enzymes. Since the addition of phosphate groups greatly increases the activity of certain enzymes, P_{fr} might thus indirectly control the action of many cellular catalysts, and thereby the mass biochemistry of the cell. The evidence for this theory is not convincing, and most recent data cast doubt on P_{fr}'s ability to attach phosphate groups. It is still possible, of course, that P_{fr} phytochrome carries out some other enzymatic activity that has not yet been discovered.

Phytochrome and membrane permeability

In the virtual explosion of experimentation that followed the isolation of phytochrome in the late 1950s and the introduction of a convenient, commercially available instrument for measuring the pigment in tissues, many sought to measure the levels of P_{fr} phytochrome before and after light was administered to control flowering, seed germination, or etiolation. Some of these experiments were reasonably successful, but a few noteworthy exceptions forced scientists to reappraise the significance of spectral measurements and the role of phytochrome. In etiolated pea stems, for example, red light converts P_r to P_{fr} and inhibits elongation, while far-red converts P_{fr} to P_r and promotes elongation, as expected. Yet, far-red light continues to promote elongation even when there is no longer any spectroscopically detectable P_{fr}! Corn seedlings display another paradox: weak red light that promotes bending toward light and gravity produces no detectable change in P_{fr}, although it should.

These and other paradoxes have been logically resolved by assuming that there are at least two pools of phytochrome in the cell. One very small pool, assumed to be physiologically active, makes no large contribution to the spectrum, while another, larger pool is inactive but absorbs light heavily. Since other evidence has shown that a small part of the cell's phytochrome is attached to the membrane, it has been proposed that only such membrane-bound phytochrome is active physiologically. The bulk of the phytochrome would constitute an inactive reservoir.

If all active P_{fr} is indeed bound to membranes, then it might act by altering the permeability to specific molecules of the plasma membrane or internal membranes of the cell. The signal would then be amplified because a few P$_{fr}$ molecules would control the passage (or nonpassage) of many ions or molecules into the cell or a particular cellular compartment. The altered chemical milieu could then result in altered physiological behavior.

There is considerable evidence for the operation of this chain of events in widely divergent groups of plants. In the alga *Mougeotia,* for example, probes with a tiny beam of light reveal that phytochrome lies not in the chloroplast itself, but rather at the periphery of the cell, in or near the plasma membrane. Each cylindrical cell of *Mougeotia* contains a single flat, platelike chloroplast that can rotate in response to changing light conditions. The rotation is controlled by phytochrome, in that the edge of the chloroplast always assumes a position as close as possible to the region of the cell's periphery having the least P$_{fr}$. Thus, red light causes the chloroplast to lie in a plane perpendicular to incoming radiation, whereas far-red causes rotation to a plane parallel to the light. Each wavelength reverses the effects of the other.

Phytochrome seems fixed in one orientation in its location at the cell's periphery, since red and far-red plane-polarized light can be perceived only when presented at certain orientations, but not at others. (If phytochrome molecules were simply in solution, or free to move around, the

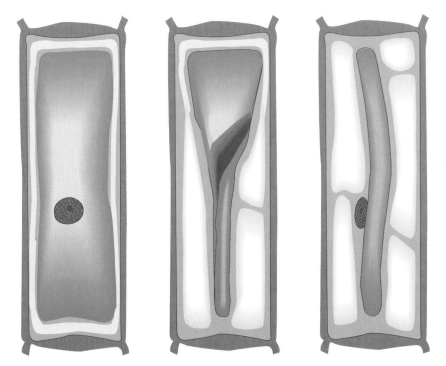

The single chloroplast in the cell of *Mougeotia* rotates so as to present its edges to the peripheral area lowest in P$_{fr}$. *Left:* platelike chloroplast in face view in red light. *Center:* bottom end of chloroplast received far-red light and has rotated. *Right:* entire chloroplast has rotated in far-red light.

plane of polarization should make no difference.) In all cases, the polarization plane of the best-perceived red light is perpendicular to that of the
best-perceived far-red light. This suggests that the orientation of part or all
of the phytochrome molecule is altered 90° by its conversion from P_r to
P_{fr}, or from P_{fr} to P_r.

The edge of the chloroplast is attached to the plasma membrane by
microfibers, and it is the contraction of these fibers that seems to make the
chloroplast rotate. Since contracted fibers are richer in calcium than un-
contracted fibers, and since fibers can be seen to contract when the concentration of calcium around them is raised, calcium seems to be necessary
to this action of microfibers. Thus, calcium may be the controlling element in chloroplast movement. Putting all these facts together, one can
construct a picture that shows phytochrome regulating the accumulation
of calcium in specific tiny regions of the cytoplasm. Phytochrome activation would allow calcium into these regions of the cell, the microfibers
would contract, and the chloroplast would rotate. Phytochrome could
exercise its control over calcium by controlling the passage of this ion
through the plasma membrane.

In *Adiantum,* the maidenhair fern, spores cannot germinate in total
darkness, but they do germinate readily after irradiation with red light.
The effects of red light are, in turn, reversed by far-red light, and vice
versa; from this behavior, we can attribute the effect of light on germination to phytochrome. Red light is effective only if sufficient calcium is
present in the medium, and red light can be shown to cause or to facilitate
the entry of calcium into the cell. Most important, spores will germinate
even when incubated in total darkness if there is present a substance that
facilitates the entry of calcium. These results again imply that the sole
function of P_{fr} is to bring calcium into the cell, presumably by some action
taken at the membrane.

In some leguminous trees with doubly compound leaves such as *Albizzia* and *Samanea,* the leaflets show "sleep movements": they are in an
open position during the day and are closed at night. Leaflets will close at
the end of a daylight period only when phytochrome is in the P_{fr} form;
thus, if far-red light is administered at twilight, leaflets will close very little
or not at all. Closure is caused mechanically by the swelling of some
specialized cells and the shrinking of others. These cells swell by taking in
water after accumulating potassium and chloride ions in response to signals from P_{fr} phytochrome and the unknown "clock" that controls the
circadian rhythm. Again, phytochrome could give the signal by acting on
the membrane so that it allows ions to enter the cell.

The leaflets seem to close in response to a well-known cascade of
biochemical events in the membrane that are also important in the re-

sponse of animal and microbial cells to stimuli such as touch, light, or hormones. A stimulus such as P_{fr} activates an enzyme in the membrane that liberates two molecules from the lipid bilayer: inositol trisphosphate (IP_3) and diacylglycerol (DAG). Following its release from the membrane, IP_3 binds to a receptor molecule on the calcium-rich endoplasmic reticulum, a network of interconnecting tubes that form the cell's internal transport system. The attachment of IP_3 to this organelle releases calcium ions into the cytoplasm. DAG remains in the membrane, where it activates an enzyme of the type called a protein kinase, which controls the activity of other enzymes by attaching phosphate groups. Enzymes activated in this way are involved in the uptake of potassium and other ions, one of the steps leading to leaf movement.

In all three of these phenomena, an early event in the chain from stimulus to response is the controlled passage of calcium ions across a membrane. This is certainly one mechanism by which phytochrome could function to control the physiology of the plant.

Switching genes on

Genes are the ultimate regulators of cell activity, for they control the synthesis of structural proteins and especially of enzymes, the workhorses of the cell. But genes themselves can be controlled, turned "on" or "off," by regulatory proteins that attach to them. External signals, such as light and temperature, can act through receptor molecules such as phytochrome to control the synthesis of regulatory proteins by other genes. These signals thus direct the plant's activity by controlling the assembly of genes functioning at any one moment.

In plants, as in animals, genes are parts of long molecules of deoxyribonucleic acid (DNA), which is composed of two strands wound in a double helix. The backbone of each helical strand is formed of molecules of the sugar deoxyribose, each linked to phosphate and all strung together in a chain. From each sugar, any one of four bases, designated A, T, G, and C for adenine, thymine, guanine, and cytosine, protrudes at right angles to the axis of the helix. The two helices are tied together by hydrogen bonds linking each opposing pair of bases; A is always paired with T and G with C. Thus, specifying the order of bases on one strand automatically specifies it on the other. This double helix provides the structural basis for both the replication of DNA and the specification of the order of amino acids in proteins.

In DNA replication, essential for the division of cells and of DNA-containing organelles such as chloroplasts, the two strands of the double helix separate, and new strands are synthesized on their surfaces with the help of appropriate enzymes, following the rules that A pairs with T and G with C. In protein synthesis, the strands also separate, but a different enzyme now intervenes to catalyze the formation of a related molecule, RNA. This molecule resembles DNA, except that the sugar is ribose and the base uracil (U) replaces thymine (T); U, like T, bonds specifically with A. The newly formed RNA goes on to regulate the formation of appropriate proteins. The step from DNA to RNA is called transcription, that from RNA to protein is called translation.

In the regulation of protein synthesis, each group of three successive bases codes for one of 20 possible amino acids. The DNA strand initiates protein synthesis by making a complementary copy of itself in the form of RNA. Part of this RNA, called messenger RNA (mRNA), contains the base sequence specifying the amino acid sequence for one protein. The mRNA directs the assembly of the protein on the surface of bodies called ribosomes by specifying the order in which particular amino acids are added to the ends of protein chains being synthesized on the ribosome surface. Each amino acid is transported to the ribosome by a specific transfer RNA (tRNA), and the tRNA inserts the amino acid in its proper position by recognizing, and pairing with, a triplet of bases on the mRNA. The result of the processes of transcription and translation is the synthesis of one specific protein, coded by a section of DNA defined as a gene. Each of the many structural and enzymatic proteins of a cell is specified by at least one gene.

Phytochrome is known to control genes important to plant functioning. For example, in photoblastic seeds, P_{fr} controls the synthesis of the enzyme amylase, essential to the digestion of starch stored in the endosperm. Without such digestion, seed germination cannot proceed. P_{fr} also increases the activity of genes controlling the synthesis of Rubisco and other enzymes involved in photosynthesis. We shall encounter other instances of the control of genes by phytochrome in later chapters.

Whenever a gene is activated, whether by phytochrome or some other controlling agent, it makes its specific mRNA. This mRNA can be extracted, separated from other mRNAs by its electrically controlled movement on a gel, and then stained by an appropriate specific probe. Finding mRNA for a particular gene is the surest sign that the gene is active, and the relative amount of RNA found can tell whether a gene is fully or only partially active. Phytochrome can both promote and inhibit the formation of mRNA by particular genes.

It is not yet clear whether phytochrome takes part in only one master reaction or in several. If phytochrome is found only in the plasma membrane, then it might activate genes indirectly, by controlling the influx of

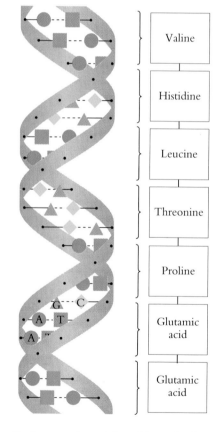

Each successive triplet of bases in the double helix of DNA specifies an amino acid.

Dark   Bright
field   field

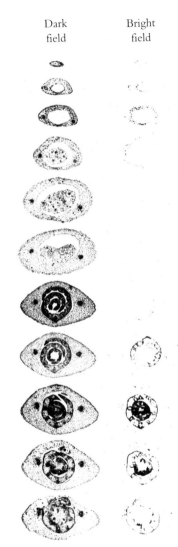

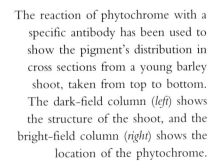

The reaction of phytochrome with a specific antibody has been used to show the pigment's distribution in cross sections from a young barley shoot, taken from top to bottom. The dark-field column (*left*) shows the structure of the shoot, and the bright-field column (*right*) shows the location of the phytochrome.

some gene-stimulatory substance. If, on the other hand, it is also present in close proximity to DNA in the nucleus, it might control gene function more directly, by combining with nucleic acid or protein components of the genetic apparatus. P_{fr} seems to promote some rapid reactions, like chloroplast rotation and leaflet closure, without gene activation. Thus, if any single reaction is the key to all phytochrome's effects, it probably occurs in the membrane.

Since phytochrome is a protein, it can act as an antigen when injected into test animals; that is, it elicits the formation of specific antibodies. These antibodies, extracted from the animal and purified, will bind with phytochrome molecules that they encounter. When, in addition, such antibodies are labeled with radioactive markers or fluorescent compounds, they can be used to locate even minute quantities of phytochrome in the plant cell. Using such antibodies, researchers have identified at least two kinds of phytochrome, differing in molecular size and antigenic specificity. Genes controlling the synthesis of each type of phytochrome have been isolated, cloned, and used to produce their phytochrome, even in foreign cells.

One gene produces a phytochrome, called Type I, that accumulates in the dark and disappears in the light, after conversion to P_{fr}. A second gene codes for Type II phytochrome, which is present in much smaller quantities than Type I but is stable in the light. The significance of these differences is not yet clear, but one theory holds that Type II phytochrome, through its own conversion to P_{fr}, controls the amount of Type I phytochrome produced. That is, in its P_{fr} form, Type II phytochrome would activate the gene controlling the synthesis of Type I phytochrome. In *Arabidopsis,* a small cruciferous weed widely used for genetic studies because of its relatively small genome (total DNA), five different genes controlling phytochrome production or action have recently been identified. Although scientists do not yet understand the meaning of the various types of phytochrome and of the genes controlling their synthesis, clearly continuing research of this kind will soon alter our concepts of phytochrome structure and function.

Phytochrome Mutants and Transgenic Plants

One way to find out what a gene does is to find an example of a malfunctioning gene of the same type and ask what went wrong. For example, the human blood disease sickle-cell anemia is caused by a faulty gene for producing hemoglobin. An analysis of the abnormal hemoglobin produced by this gene showing that it differs from normal hemoglobin by only a single amino acid has helped scientists to understand how the normal gene works. Faulty genes are produced by random events, called mutations, that alter the structure of DNA, usually by altering the sequence of bases in the DNA chain. Random mutations are relatively rare in most genes; one occurs about once every million replications of a gene.

If the seeds of a plant are exposed to certain forms of radiation or to certain chemicals, the frequency at which mutations will be introduced into the DNA is increased. As a result, some of the progeny of such "mutagenized" plants will show aberrations. In the tomato, a mutant named *aurea* is a paler green than the "wild type" plant and has longer internodes, while another mutant is deeper green and has shorter internodes. The paler plant is deficient in phytochrome, while the greener plant has an excess of this pigment. Mutants of this type have permitted us to assess the many roles of phytochrome in the plant.

Sometimes mutants contain a normal amount of phytochrome yet do not respond normally to light or its absence. These plants are aberrant because of differences in the events leading from phytochrome activation to the ultimate response. For example, the *det* (deetiolation) and *cop* (constitutively photomorphogenic) mutants of *Arabidopsis* grow to become green and short even in total darkness, an abnormality that permits their easy detection. It appears that in darkness the relevant gene in each case produces changes that in wild type plants are produced only after the conversion of P_r to P_{fr} in light. Since *det* and some *cop* mutants appear to have normal amounts of phytochrome, the genes producing these abnormalities must be controlling events after phytochrome activation. An analysis of the action of these genes should permit us to understand something about what P_{fr} does. By analyzing a collection of mutants in a single plant such as *Arabidopsis,* scientists may eventually reconstruct the chain of events between phytochrome activation and its ultimate action.

The creation of transgenic plants offers an alternative route to understanding the mechanisms of control by phytochrome. Since phytochrome genes can be identified, isolated, and cloned, it has become possible to transfer them between plants by the techniques described in Chapter 8. In this way, one can produce, for example, tobacco plants containing oat or rice phytochrome in addition to their normal phytochrome. Or, one can create plants with extra numbers of genes for their own phytochrome or plants with genes causing abnormal phytochrome action. Abnormalities in these transgenic plants will provide clues to how phytochrome acts. Already, analysis of such plants has revealed that certain phytochrome genes are expressed only in certain cells, like pollen, at particular stages of the plant's development. Evidently, the action of certain other genes must be consummated before phytochrome can produce its effect.

Wild-type (*top*) and *det* mutant (*bottom*) seedlings of *Arabidopsis* grown in dark (*left*) and light (*right*). In darkness, *det* shows the short stem and enlarged cotyledons found in the wild type only in the light.

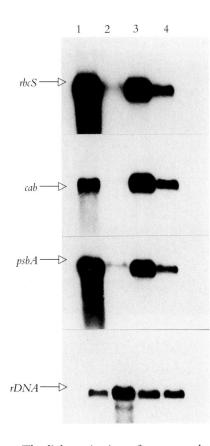

The light activation of genes can be shown by separating and staining their mRNAs. Lanes 1 and 2 show mRNAs from the "wild type" *Arabidopsis* plant grown in light and dark, respectively (note absence of mRNA for upper three genes in lane 2); lanes 3 and 4 show mRNAs from the *det* mutant. In darkness, *det* partly "turns on" the *rbcS, cab,* and *psbA* genes that need light in the wild type.

Phytochrome research is currently one of the most active areas in all of plant physiology and molecular biology. Now that the molecule has been purified, isolated, and structurally analyzed, and now that we have come to understand some of the genes controlling its synthesis, the mechanism through which it controls so many aspects of plant behavior will probably be revealed in the near future.

The Ecological Significance of Phytochrome

Plants growing in the wild compete for many necessities, including space. Frequently, many seeds will fall within a small area, and each will germinate to produce a plant with spreading foliage that may block out light to shorter plants below. Because access to light affects the growth, photosynthetic capacity, and very survival of seedlings, they will compete vigorously for a place in the sun. In this highly competitive environment, phytochrome can play a crucial role.

The shade of a leafy canopy screens out precisely those wavelengths of light to which phytochrome is most sensitive. Chlorophyll absorbs heavily in two regions of the spectrum, the red and the blue. The red absorption peak at about 665 nm coincides almost exactly with the absorption peak for P_r phytochrome. Consequently, a single green leaf will serve as a filter, permitting the passage of far-red light that converts P_{fr} to P_r, but absorbing the red light that converts P_r to P_{fr}. Thus a plant growing under a leaf or a canopy of leaves receives far-red light in profusion, but not much red light. Like an etiolated plant, it develops a deficiency of P_{fr} and an abundance of P_r. In such a plant, the stem extends greatly while leaf growth diminishes. The elongation of the stem may carry the previously shaded plant to a position above the plant that had been shading it, and its leaves will now in turn shade that of its competitor.

A leaf may shade other leaves within a single plant. It is well known that the "sun leaves" of many plants are thicker and greener than "shade leaves" and are different in internal structure and chemistry. The stems leading to sun leaves are also thicker, and they are shorter as well. The elongated petiole of a shaded leaf can bring the blade into the sunlight, where the synthesis of chlorophyll will be stimulated and photosynthetic activity raised. These differences in sun and shade leaves are at least partly the result of the ratio P_{fr}/P_r to which the developing leaf is exposed. Even the reflected light from close neighbor plants can shorten and thicken the stems and green the leaves of neighboring plants, owing in part to the activation of phytochrome.

A transgenic tomato overexpresses phytochrome (*left*) compared with a "wild-type" control (*right*). Note the greatly stunted stature of the transgenic tomato.

Within a dense community of plants, phytochrome may have other ecological effects. Consider a forest floor, rich in far-red light, on which seeds have fallen. If the seeds lie dormant because of a deficiency of P_{fr}, they will remain ungerminated until irradiated with sunlight, which is rich in red wavelengths. Should a part or all of the canopy be cleared away, as by a forest fire, a virtual explosion of germinating seeds will fill the clearing.

The importance of the pigment phytochrome in plant development can hardly be overestimated. The more we look, the more we find that it influences almost all phases of plant growth and development, from seed germination to senescence. Yet, we have known about its existence for only several decades, and the details of its structure and probable mechanism of action are only now becoming clear. Its history illustrates the interplay of chance, systematic investigation, and occasional genius that characterizes most scientific discoveries.

While much remains to be learned about the intermediate steps between light reception and ultimate physiological response, phytochrome clearly does not act alone. Hormones synthesized by the plant can also induce bolting, flowering, and other responses identical to those initiated by phytochrome. These chemical substances, to be discussed next, sometimes appear to act as messengers between phytochrome and the plant's response.

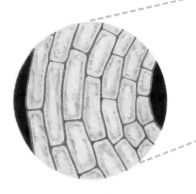

The multicellular plant body, often meters long, presents a special problem: How are the growth and activity of its parts to be coordinated? To be well formed and to function harmoniously with the rest of the plant, each organ must be in some kind of communication with the other plant parts; lacking such interaction, each organ could grow independently of the others, resulting perhaps in a malfunctioning organism. Yet the green plant lacks a nervous system or any analogous set of cells specialized for the rapid transmission of long-distance signals. Its organs communicate almost exclusively through a much slower mechanism: the movement of microquantities of small molecules called hormones. These substances are commonly termed chemical messengers, because their role depends on their moving from one location, where they are synthesized, to another location, where they act. Only small quantities of hormones are needed to produce large effects in the cell because, like phytochrome, their action is amplified through processes related to membranes, enzymes, or genes.

THREE

Growth and
Chemical Signals

The Discovery of Plant Hormones

Groundwork for the existence of plant hormones was laid in the 1880s by Charles Darwin and his son Francis, working with the well-known but poorly understood curvature of plants toward sunlight. This phenomenon was originally known as heliotropism, indicating a curvature toward the sun, but is now called phototropism, because light energy from any source, not only the sun, induces the curvature. In general, tropisms are curvatures of plant parts toward or away from an asymmetrical stimulus; they are caused by a difference in the rate of growth on the two sides of the plant axis. Plants also respond tropistically to gravity (gravitropism), water (hydrotropism), and touch (thigmotropism).

The Darwins had previously noted that the tips of all plant organs describe random oscillatory patterns called circumnutations as they grow. They had studied such movements, especially in the stems of various herbs and vines and in the cylindrical leaf sheaths (coleoptiles) of grass seedlings, and wondered about their function. They discovered that circumnutation is the result of random displacements of the growing tip from the vertical axis; the tip moves in a spiral pattern as it elongates, without net displacement from the vertical. Unilateral light alters this pattern, so that the circumnutating organ sustains a net curvature toward the direction of the incoming light beam; gravity functions in the same way. Thus, circumnutation is important because it provides the basic mechanism that makes tropistic curvature possible.

By placing opaque barriers over parts of the delicate seedlings of canary grass and, later, oats, and then illuminating only a small exposed area,

Left: Charles Darwin observed these random patterns of circumnutation during a period of 10 hours 45 minutes. *Right:* Under the influence of an asymmetric light field introduced where the pattern straightens out, random circumnutation is converted into a directional curvature.

the Darwins determined that the tip of the growing grass coleoptile is the area by far the most sensitive to the impinging light. Yet, the region of curvature is centered not at the tip but a centimeter or more below, at a distance of many cell diameters from the tip. In their 1880 book *The Power of Movement in Plants,* the Darwins hypothesized that some influence from the tip, probably a chemical substance, moved down the coleoptile to regulate the comparative elongation rates on the two sides of the organ, and thus its curvature. This suggestion was not greeted enthusiastically by the era's pundits of plant physiology, in part because the Darwins were unknown outsiders in the field. Charles Darwin had attained notoriety, but not yet acceptance, for his theory of evolution. Unfortunately, the tendency to resist the sometimes heretical views of unknown outsiders persists in the scientific establishment even today. Although the Darwins' hypothesis was rejected by the establishment, it did lead to experiments by others that ultimately resulted in the discovery of a growth-regulating substance such as the Darwins had proposed.

Starting about 1910, a Danish plant physiologist, Peter Boysen-Jensen, repeated the Darwin experiments and found that if the tip were cut off and merely replaced on the coleoptile stump, the stump still responded. Thus, the tropistic stimulus could pass from the light-stimulated tip to the curving region of the coleoptile across a wound gap. Arpad Páal, in Hungary, then showed that the tropistic stimulus could cross even a layer of gelatin, but was blocked by a barrier of mica, platinum foil, or cocoa butter. This suggested that the stimulus passing from the tip to the lower parts of the coleoptile was a water-soluble chemical substance. Páal also showed that if the coleoptile tip was replaced on only one side of the cut stump, curvature could occur even without light. Hans Söding, a German plant physiologist, then demonstrated that this mobile substance promoted growth: removing the tip of the coleoptile diminished growth of the lower regions, whereas replacing it symmetrically restored growth at the original rate. These experiments suggested that a water-soluble growth promoter moved from the tip to influence elongation of the tissue below. If its distribution was symmetrical, growth was straight; if its distribution was asymmetrical, the coleoptile curved.

This theory was logical, but was it biological? For a while, it seemed not. Many able biochemists of the era attempted to obtain the then hypothetical growth promoter by extracting it from cut-up or ground-up coleoptiles with various solvents, such as alcohol or ether. Despite a few reports of success, most extracts inhibited growth rather than stimulated it, and no convincing pattern emerged. It remained for the simplest, most naive approach to bring the search to a successful conclusion.

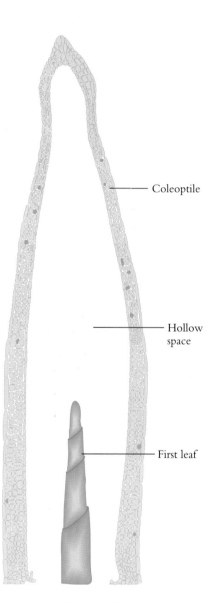

Coleoptile

Hollow space

First leaf

The coleoptile, or leaf sheath, is a hollow cylinder several cells thick that grows entirely by cell enlargement. It encloses the coiled first true leaf, which eventually grows through the tip of the coleoptile.

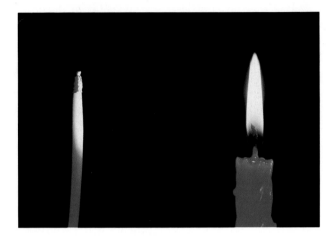

A coleoptile curves toward the light of a candle (*top, left*). If the extreme tip is shaded from the light, curvature does not occur, even though the rest of the coleoptile is illuminated (*top, right*). If a barrier prevents illumination of all of the coleoptile except the tip, curvature still occurs (*bottom, left*).

A full 50 years after Darwin's original suggestion, the definitive experiment, proving the existence of a growth hormone in the coleoptiles of grass seedlings, was finally performed. Working at the University of Utrecht in the Netherlands, a young graduate student named Frits Went showed that excised coleoptile tips could diffuse a growth-promoting material into a watery gel made of agar, a gelatinlike material derived from seaweed. When the agar block containing such material was placed symmetrically on the cut stump of a decapitated coleoptile, it produced the growth-promoting effect of the tip; if placed asymmetrically on the cut stump, it caused curvature. Furthermore, when light was shone onto one side of a tip, its shaded and illuminated sides produced different quantities of the growth-promoting substance. The asymmetrically illuminated tips were excised and split longitudinally into lighted and shaded halves, and each half was permitted to diffuse its hormone into an agar block. As Went expected, the shaded side yielded twice as much hormone as the lighted side. The two agar blocks would induce curvature, as the tips had, if placed on opposite sides of the top of a decapitated coleoptile.

The explanation of curvature was now established: somehow, light hitting the tip from one side causes that side to export less growth hormone than the other. As a result of this asymmetry in the export of the growth hormone, named *auxin* from a Greek word meaning growth, cells lower down in the coleoptile elongate at different rates, ultimately resulting in curvature. This theory, formulated jointly by Went and a Russian physiologist named Nicolai Cholodny, has served for years as the explanation of most plant tropisms. Currently, the Went-Cholodny theory is under serious challenge, a subject that we will explore in the next chapter.

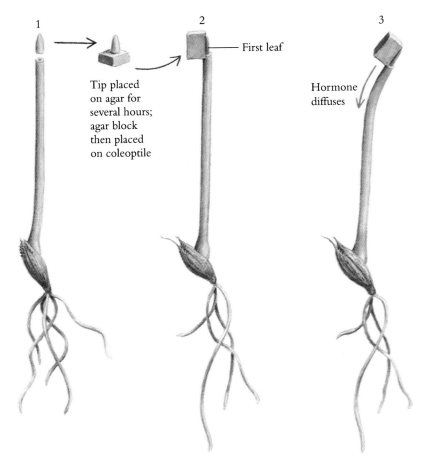

1

2

3

First leaf

Tip placed
on agar for
several hours;
agar block
then placed
on coleoptile

Hormone
diffuses

Went's experiment proved the existence of a growth-promoting hormone by showing that a coleoptile would curve when exposed to an asymmetric application of an agar block containing the hormone.

Went was able to show that the angle of curvature induced in a decapitated oat (*Avena*) coleoptile by an agar block placed on one side of the cut stump is proportional to the amount of auxin in the agar. The angle of curvature then became the basis of a bioassay for the hormone. A bioassay is a procedure for estimating the amount of a physiologically active substance by measuring the response of an organism to various concentrations of that substance. Bioassays are commonly used before the chemical nature of the substance is known and chemical or physical assays have been designed.

Bioassays can be useful in establishing the characteristics of the substance and in isolating it from the other components of plant tissue. For example, suppose that a plant tissue shows hormone activity from an unknown substance that will enter into solution in water, along with all the other water-soluble components of the tissue. If the solvent ether is added to the aqueous solution containing the active substance, a layer will form in the liquid containing all the substances present that dissolve in that solvent. Both the ether layer and the residual water layer can then be bioassayed for hormonal activity. If it is found in the ether layer but not in the water, it can now be transferred to another solvent, such as alcohol. At each transfer, the active substance is separated from other substances; ultimately, a highly purified fraction may produce crystals of the active substance, or chemists may be able to deduce the nature of the active molecule in other ways. Although Went discovered much about the properties of auxin, it took the collaboration of biochemically inclined colleagues at Utrecht to finally isolate and identify it.

Because auxin is present in vanishingly small quantities in the oat coleoptile, its isolation from that organ would have been a daunting task, even for skilled chemists of the day. A more abundant source of the substance was needed. Although it might have seemed logical to investi-

The curvature of the coleoptile depends on the amount of auxin in the agar block. The angle of curvature thus serves as a bioassay for the hormone.

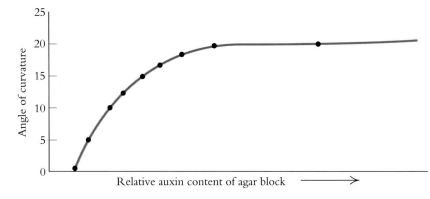

gate other plant materials first, it was not long before Went's colleagues, the chemists Fritz Kögl and Arie Haagen-Smit, discovered that substances promoting the curvature of coleoptiles could be found in many natural sources in addition to plants, including human urine. A basic food nutrient ingested in the diet undergoes metabolic transformation to become the active substance found in urine, and large quantities can be readily extracted for chemical identification. By extracting repeated fractions and assaying them using the oat-coleoptile curvature test, Kögl and Haagen-Smit were able to show that the active substance in urine is a simple molecule, indole-3-acetic acid, known colloquially as IAA. Later on, other investigators found that IAA also exists in plants and is identical with the stimulus regulating the growth and tropistic curvature of the oat coleoptile. IAA was thus the first plant hormone to be identified.

In the decades since auxin was identified, four other plant hormones have become well characterized; they are cytokinin, gibberellin, abscisic acid, and ethylene. As discussed in Chapter 2, there is also strong evidence for the existence of a floral hormone, although such a hormone has not yet been isolated, despite decades of intense work. The list of hormones may not end here, because, as research continues, evidence for new hormonal regulatory systems may be discovered.

It is believed that the target organ for each hormone contains a receptor system, probably a protein, to which the hormone binds. This union creates a change in the form of the protein, and that change in form ultimately leads to physiological changes of a magnitude dependent on the quantity of hormone received. In this way, chemical signals regulate most of the plant's vital activities, including germination, cell division, cell extension, the relative growth rates of different organs, the onset of flowering, the ripening of fruits, seed and bud dormancy, and senescence. Thus, from germination to death, the plant's life cycle is regulated by transportable chemical signals.

The Roles of Auxin

The availability of relatively large quantities of pure IAA made many new experiments possible, and experimenters all over the world entered the new field of plant growth hormones. It soon became clear that auxin controls many processes in addition to the curvature of coleoptiles toward light. Most important, it appears to regulate one of the most basic of all plant growth processes, the elongation of the young cube-shaped cells produced by meristems to mature, brick-shaped cells, which occurs in all growing plant organs.

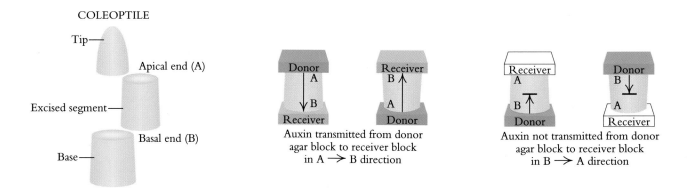

COLEOPTILE

Tip

Apical end (A)

Excised segment

Basal end (B)

Base

Auxin transmitted from donor
agar block to receiver block
in A → B direction

Auxin not transmitted from donor
agar block to receiver block
in B → A direction

Auxin movement is polar, from apex to base of the coleoptile or stem. When auxin is applied to the apical (A) end of a piece of coleoptile, it moves through the tissue—even against gravity. When applied to the basal end (B), it will not move, even in the direction of gravity's pull.

IAA is somehow also involved in maintaining the "polarity," or top versus bottom organization, of the plant body. A poorly understood phenomenon, polarity depends on the fact that auxin moves from apex to base of the plant but not easily in the reverse direction. It appears that auxin is moved from cell to cell by an auxin pump in the membrane at the lower end of each cell; this pump can transport auxin in one direction only, from the bottom of one cell to the top of its lower neighbor. Once across the membrane of the lower cell, auxin diffuses down to the membrane at the cell's base, through which it is again pumped downward. When auxin reaches its target cells just below the meristem, it regulates the rate at which they absorb water and expand. Also, in combination with other substances, it regulates the rate of meristematic cell division.

While IAA is found widely in the plant kingdom, it is only one of a group of related compounds called auxins. All auxins have similar effects on plant growth: they all pass the oat-coleoptile curvature test and they all promote the growth of excised cylinders of oat coleoptiles and various other cylindrical plant organs. Other auxins having most of the physiological effects of IAA while differing from it in basic chemistry can be synthesized in the laboratory. For example, α-naphthyl acetic acid (NAA) is about as active as IAA; the same can be said for 2,4-dichlorophenoxyacetic acid (2,4-D), a common herbicide. While IAA, NAA, and 2,4-D are all active auxins, only IAA can be referred to as a plant hormone, because it alone occurs naturally. It appears that all auxins have in common some sort of an aromatic ring structure (a ring of six carbons containing alternating single and double bonds, as in benzene) to which is attached a side chain of several carbon atoms terminating in either an acidic carboxyl group (COOH) or a chemical precursor of the carboxyl group.

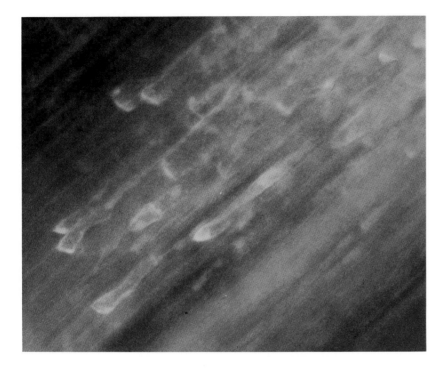

Fluorescent antibodies highlight the location of presumptive polar auxin pumps at the base of cells engaged in the polar transport of auxin.

Most plants seem to produce approximately enough auxin to grow at an optimal rate. However, in a few cases, such as some genetically dwarf pea mutants, it is possible to promote growth by applying active auxin molecules to the shoot. Applied auxin can also affect the location of growth on a plant. In general, the apical bud on a leafy shoot sprouts in preference to the lower buds, the degree of dominance being genetically determined. If the apical bud is removed, then one of the lower buds will

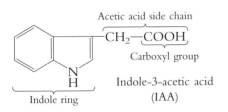

Indole-3-acetic acid
(IAA)

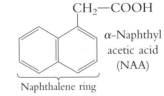

α-Naphthyl acetic acid
(NAA)

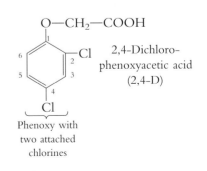

2,4-Dichloro-phenoxyacetic acid
(2,4-D)

Apical dominance is caused by the downward movement of auxin from the tip (1). Removal of the apex permits a lower bud to sprout (2), but replacement of the tip by an agar block containing auxin prevents this sprouting (3).

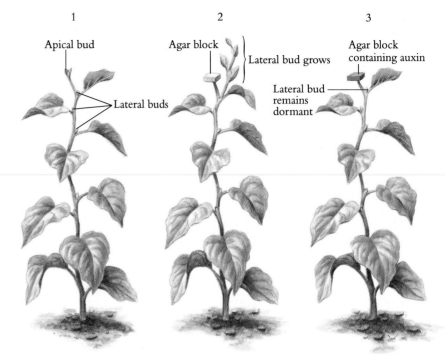

sprout readily, but if a paste containing auxin is applied to the cut surface of the stem, the lower buds will remain in quiescent state. Thus, it appears that auxin, diffusing downward from the stem apex, normally keeps lateral buds in check; auxin thereby serves as a growth retardant for some types of cells while promoting the growth of others. The basis for this dual action is not completely understood, but as we shall see below, it may involve auxin's control of the production of another hormone, ethylene.

Auxin moving down the stem eventually reaches the root, where much of it is metabolically destroyed or changed to an inactive form. If a stem is cut at its base, however, the downward-moving auxin accumulates just above the cut. The cutting injury induces cells near the cut surface to divide, sometimes producing a swelling at the base of the cut stem. The swelling may in turn be followed by the emergence of a set of *adventitious* roots, which are roots that arise from unusual locations. In commercial practice, cuttings are dipped in auxin to induce adventitious root formation. The adventitious roots are produced when auxin induces cell divisions, either in undifferentiated callus cells produced just above the cut or in a layer of cells called the pericycle.

To picture the pericycle, recall that the seedling root growing down into the soil consists of a central core of relatively thick-walled, elongated vascular cells, surrounded by the thinner-walled, more cube-shaped cells

Adventitious prop roots commonly grow at the base of a corn stem.

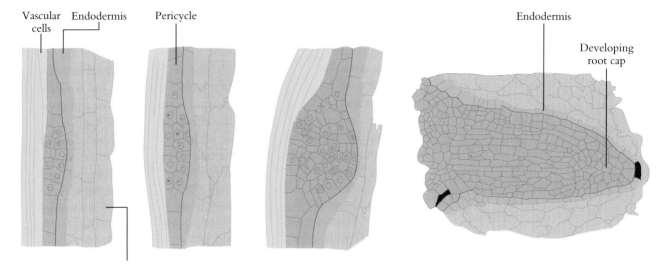

Vascular cells | Endodermis | Pericycle | Endodermis | Developing root cap | Cortex

A mass of dividing cells in the pericycle grows through cortex, developing into a branch root.

of the cortex, which are in turn enveloped by the flattened cells of the epidermis. Eventually, the central vascular cylinder becomes surrounded by a cylinder of specialized endodermal cells, which effectively seal the vasculature off from easy contact with other tissues. Inside the endodermis lies a group of cells, the pericycle, whose division can give rise to branch roots. When pericycle cells are stimulated to divide, the mass of new cells forms an organized meristem that grows its way through the cortex and epidermis, eventually emerging at the surface. Because the pericycle layer is the normal source of branch roots, as well as the source of auxin-stimulated adventitious roots, the conclusion suggests itself that the emergence of branch roots normally depends in some measure on an auxin supply to pericycle cells.

Auxin sprayed on the unpollinated ovaries of certain flowers will cause the ovaries to develop into parthenocarpic fruit, lacking seeds. Growers habitually apply auxin to stimulate fruit development in certain crop plants like tomato, which are not easily pollinated when grown in the greenhouse. At a later stage of the development of certain fruits, application of high concentrations of auxin may also lead to ripening and premature senescence. As we shall see later, this is an indirect effect arising because auxin stimulates the production of gaseous ethylene, which is the true ripening hormone. This is an example of the interaction of hormones in controlling the plant's life cycle.

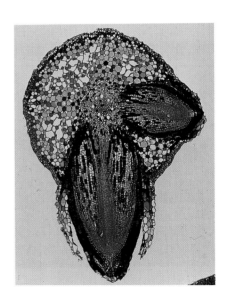

A branch root emerges from the primary root cylinder of a willow, and another is about to break through.

The large, thin-walled cells of tobacco pith do not divide in the plant
(*left and upper right*), but divide vigorously when supplied with auxin and
cytokinin in culture (*center*). The tissue does not show growth until the
new cells enlarge.

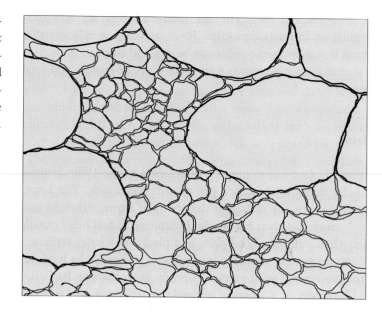

produces cytokinin, which regulates cell division and other aspects of
stem growth. Since the quantities of auxin and cytokinin exported are
roughly dependent on the bulk of the stem and root tissue, respectively,
this arrangement guarantees a coordinated pattern of development. The
activity of mobile chemical messengers not only ensures a desirable ratio
of root mass to stem mass, it also communicates other information, such as
the direction of gravity, light, or water, which may determine the rate and
direction of stem or root growth.

Cytokinins also oppose the inhibitory action of auxin on the growth
of lateral buds. It appears that whether or not a bud sprouts on a stem
depends on the ratio of auxin coming down from above and cytokinin
coming up from below. Counterflowing streams of auxin and cytokinin
thus determine the fate of a lateral bud at any node on the stem. Buds
located at nodes closest to the apex of the stem, where auxin is produced,
would tend to be under the influence of auxin rather than cytokinin, and
would remain dormant. The dominance of the apical bud permits the
upward growth of the stem to proceed vigorously. If, however, the apex
of the stem should be injured and the downward flow of auxin interrupted, one of the lateral buds, now under the influence of a dominant
cytokinin stream from the roots, will sprout and take over as the new
dominant apical bud. This arrangement gives the plant great resilience in
overcoming mechanical injury. Buds located at nodes near the root sys-

tem are less affected by the distant stem apex and, in some plants, may sprout even when the stem apex remains intact.

The interaction of auxin and cytokinin is well illustrated by long-term cultures of excised tobacco pith. In the presence of high concentrations of both auxin and cytokinin, cells divide vigorously to form an undifferentiated callus mass. Dropping the cytokinin level somewhat leads to root formation, whereas lowering the auxin level somewhat leads to bud formation. Thus, assuming the concentration of each hormone is high enough, when the auxin to cytokinin ratio is low, buds form; when the ratio is high, roots form; and when the ratio is median, callus forms. This information is useful in the commercial micropropagation of certain desirable plants.

Cytokinins are known to promote protein synthesis, but whether this ability is related to their true function is not entirely clear. One natural cytokinin, isopentenyl adenine (IPA), is present in molecules of some species of transfer RNA. However, when one feeds an active cytokinin to a plant, the entire molecule is not incorporated into tRNA. Rather, the isopentenyl group is removed and only the adenine nucleus is incorporated into tRNA; later, the isopentenyl group is restored. This puzzle makes us wonder whether the presence of cytokinins in tRNA is at all related to their growth regulatory action.

These calli have produced roots under the influence of a high concentration of auxin and a lower concentration of cytokinin.

Although the original assay for cytokinin activity involved stimulating cell division in excised pieces of tobacco pith, it was later shown that cytokinins have other effects on plants. The most interesting of these is the prevention of senescence of certain excised leaves or leaf parts. Thus, when cytokinins are applied to restricted regions of leaves, islands of green remain in a developing sea of yellow caused by chlorophyll disappearance. Cytokinins are sometimes used commercially to maintain the greenness of excised plant parts, such as cut flowers. Their use in edible crops such as broccoli, however, is still not permitted in the United States, because any chemical like cytokinin that resembles a nucleic acid component is automatically a suspected carcinogen.

Gibberellins

In the late nineteenth century, Japanese rice farmers noted extraordinarily tall seedlings rising at intervals in fields of otherwise uniform plants. Hoping that these tall plants might constitute a strain of giant rice, the farmers watched the seedlings in the expectation that they would flower and produce grain to be used as seed. The seedlings never reached sexual maturity, however; instead the stems grew unusually rapidly, then died before flowering. For this reason, the Japanese farmers named this phenomenon the "foolish seedling" or *Bakanae* disease.

In the 1920s, a Japanese botanist named Eiichi Kurosawa discovered that these foolish seedlings were all infected by a fungus, *Gibberella fujikuroi.* If the spores of the fungus were transferred from an infected to an uninfected plant, the latter became diseased or hyperelongated. If the spores were placed in an artificial medium and permitted to grow into a fungal mass, then the liquid in the culture also contained the active principle producing hyperelongation of normal seedlings. From this liquid, Japanese scientists were able to isolate and identify the active material, which they named gibberellic acid. The chemical structure originally proposed by the Japanese was somewhat in error, and in the 1950s a group of British scientists at Imperial Chemical Industries succeeded in extracting and identifying gibberellins from other sources. Since then, more than 80 similar molecules, collectively called *gibberellins,* have been found in higher plants. Structurally, the gibberellin molecules are roughly analogous to the steroid group of animal hormones, and since they move from their site of synthesis to a distant site of physiological action, they qualify as plant hormones.

When gibberellins are applied to certain genetically dwarf plants, the plants grow to normal height, whereas normal plants sprayed with gibberellins fail to respond at all or respond only very little. Gibberellin thus seems to be essential for normal growth. It was therefore originally reasoned that dwarfness must result from inadequate gibberellin content. Unfortunately, chemical analysis does not always support that conclusion, since some dwarf plants show normal gibberellin content. However, these dwarfs react less well to applied gibberellin than do tall plants. Some other substance involved in how gibberellin produces its effect may instead be deficient in dwarf plants. That substance could be a protein receptor in the target cell, to which gibberellin must attach before it acts. Some plants do not manifest their dwarfness in darkness; only after phytochrome transformation from P_r to P_{fr} does the dwarfness genotype show up. In such photodwarfs, P_{fr} must somehow control either the level of gibberellin or plant response to gibberellin.

Of the more than 80 different gibberellins known, the most commonly used in experimentation is the fungal product gibberellic acid or GA_3. Other gibberellins differ from GA_3 in the number and position of chemical groups such as methyl (CH_3) and hydroxyl (OH) added to the molecule's basic ring-shaped nucleus. Each of these variously modified gibberellins, derived from a chemical precursor common to all gibberellins, is found preferentially in a particular plant or plant family, and each elicits somewhat different physiological effects when applied to different plant species or to different organs within the same plant. For example, gibberellins produce special effects on many biennial plants that show the rosette habit of growth. In such plants, virtually no stem elongation occurs during the first year, and the only growth is a whorl of leaves at ground level, the rosette form in which many biennial plants overwinter.

When gibberellins are applied to rosette plants, certain cells in their stems are stimulated to divide and then to elongate rapidly and massively. This phenomenon, called bolting, is usually followed by the formation of flowers, but at low levels of gibberellin application, only bolting occurs. This result implies that higher levels of gibberellin are required for flowering than for bolting. Because bolting and flowering in biennial plants are also induced by low-temperature vernalization treatments followed by photoperiodic long days (see Chapter 2), some people have hypothesized that gibberellins are produced in the plant after such treatments, and that they must be closely related to or identical with the floral hormone. But while gibberellins can induce flowering in many long-day plants, they are inactive in short-day plants. Since studies with grafted plants show that a

A dwarf tomato (*left*) more than doubles in height after being sprayed with gibberellin (*right*).

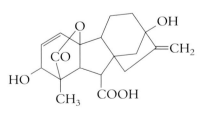

Gibberellic acid (GA$_3$)

flowering promoter can be successfully transmitted between the two types, many scientists believe that the floral hormone must either be identical in all plants, or be easily produced by conversion of a precursor molecule from the graft partner. It is still possible that the conversion of one gibberellin to another could provide the answer to this puzzle, but more experimentation will be needed to prove this point.

Gibberellins cause certain dormant seeds to germinate promptly, and in the absence of gibberellin or an appropriate environmental treatment, such seeds may remain dormant for long periods of time. As we have seen, some dormant seeds are activated by the red light absorbed by phytochrome and others are activated by low temperatures. In both types of seeds, gibberellins induce germination by accelerating the use of reserves stored in the endosperm in the form of starch grains, fat globules, or protein granules. Such reserves must be digested by specific enzymes before they can be utilized for growth. Gibberellin stimulates the production of the required enzymes and in doing so initiates the germination process; cytokinins and even some nonhormonal substances can produce similar effects on particular seeds.

The activation of germination in cereal grains is the only physiological process induced by gibberellin that has been subjected to detailed molecular analysis. In seeds deprived of their embryo, starch in the endosperm is not digested, because α-amylase is not produced. If such embryoless seeds are supplied with gibberellin from the outside, the activity of the starch-degrading enzyme α-amylase increases rapidly. Amylase cannot be produced, however, when substances interfering with the synthesis of nucleic acids and proteins have been applied to the seeds. Thus, gibberellin must produce its effect by activating the genes that code for α-amylase.

During the normal course of germination, the embryo takes in water, and in response its growing points produce minute amounts of gibberellins. These gibberellins migrate to the aleurone layer, a group of cells just within the fused fruit and seed coats of the grains; these cells contain starch and protein granules of various kinds and also an active enzyme-synthesizing machinery. Gibberellin quickly activates certain genes, probably through binding to a receptor protein. These genes code for enzymes such as α-amylase, which are then secreted into the adjacent endosperm cells, where they digest starch and other stored food, making it available to the growing points for cell division, cell elongation, and other aspects of growth.

When the effects of gibberellin on hyperelongation of cereals and other grasses were first noted, it was anticipated that by applying tiny quantities of this hormone in the field, farmers might accelerate crop

After repeated sprays of gibberellin, a rosette cabbage plant bolts and flowers. Controls at left were sprayed with water.

growth and obtain greater yields at harvest time. The correct chemical characterization of these compounds in the 1950s by scientists at Imperial Chemical Industries led to a patent for a new class of plant growth regulators and to a projected large-scale commercial agricultural enterprise to exploit their effects. Unfortunately, this activity has not paid off well. Gibberellin does promote growth early in development, but later on the untreated controls catch up with the gibberellin-stimulated plants, so that the two groups produce approximately equal yields.

Gibberellins have found important economic use in the production of both beer and wine. During beer making, gibberellin is applied to promote the germination of barley during the malting process; its use ensures a more complete conversion of starch to sugars, which are then fermented by the yeast. When applied to grape vines, gibberellin causes an extension of the stalks bearing each developing fruit in a cluster. The fruits have more room to expand and the overall yield increases. Through such unexpected paths have Kurosawa and his successors benefited generations of beer and wine aficionados!

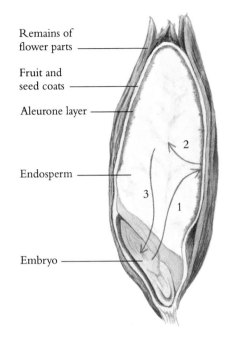

Remains of flower parts

Fruit and seed coats

Aleurone layer

Endosperm

Embryo

Gibberellin secreted from the embryo (1) stimulates the aleurone to form amylase, which is secreted into the endosperm (2), where it makes possible the digestion of stored starch. The sugars produced by starch digestion are then absorbed by the embryo (3) and used for growth.

Abscisic Acid

The hormone abscisic acid was discovered independently and almost simultaneously by plant physiologists in the United States and Great Britain while working on different problems. At the University of California, Davis, Frederick Addicott and his colleagues discovered that extracts of aging cotton bolls and other senescent tissues promoted the *abscission,* or shedding, of leaves. They developed a bioassay that measured the effects of extracts on the abscission of a test petiole under the stress of a standard applied weight or physical blow. From the most active extracts, they were able to isolate crystals of a chemical that strongly promoted the detachment of cotton petioles from stems; they named this substance *abscisic acid.*

In Great Britain, Philip Wareing and his colleagues at the University College of Wales, Aberystwyth, were studying the formation of dormant winter buds in various woody species under the influence of short days. From these buds, they extracted a chemical that dramatically slowed stem growth and caused other buds to become dormant. This chemical turned out to be identical with Addicott's. Because Addicott's report was published slightly earlier than Wareing's, the chemical was officially named abscisic acid, now designated as ABA. It is clear, however, that ABA is more closely associated with the induction of dormancy than with abscis-

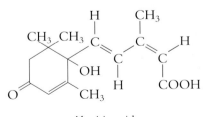

Abscisic acid

sion, which is largely regulated by ethylene. A more appropriate name, therefore, might be *dormic acid*.

ABA provides a chemical means for delaying germination until spring, used by some plants instead of the mechanical tricks and photosensitivity described in Chapter 2. As a seed ripens on the mother plant, it usually develops a high concentration of ABA. The hormone somehow prevents the "turning on" of the metabolic machinery required for the onset of growth; the seed is blocked in its development just as effectively as if it were deprived of water or oxygen by a physical barrier. When the seed is wetted, however, other chemical processes are initiated, through which ABA is broken down to innocuous materials. When the concentration of ABA becomes low enough, germination can proceed. Thus, the initial concentration of ABA and the rate of its breakdown act as a kind of chemical "hour glass" that determines the timing of seed germination.

ABA begins forming in the leaves when the plant senses that daylength is becoming progressively shorter. The hormone is transported into the seeds, where it accumulates and prevents germination. During its dormancy, the seed metabolizes away the accumulated ABA; when the hormone falls to a permissive level, germination begins. Usually the seed starts to sprout just when the seasonal climate is best for the plant to begin active vegetative growth. Clearly, the onset of ABA production is precisely timed, and the amount of ABA stored in the seed genetically determined to be just sufficient to last until the local climate is right. The timing and rate of ABA destruction must be similarly controlled.

In some mutants of corn and other plants, the seed does not store enough ABA to restrain germination, and the seed starts growing right on the mother plant. This mutation, termed vivipary, hinders the propagation of corn because the viviparous seeds are too easily damaged to permit their use in conventional agriculture. For other plants, such as the mangrove that colonizes the water's edge where tropical rivers flow into the ocean, viviparous germination is advantageous. The heavy, mature seedlings fall off the mother plant root first; their roots bury into the soft mud, and the seedlings are ready to commence growth immediately. If an ungerminated seed were to be released into such an environment, it might be eaten by a member of the abundant animal populations of the mangrove swamp or it might not readily find a convenient spot in which to germinate.

As Wareing's studies suggested, the formation of ABA and similar substances in response to declining daylength may also control the formation of special winter buds on trees, as well as the winter dormancy of woody twigs in general. As with seeds, the accumulated ABA must be

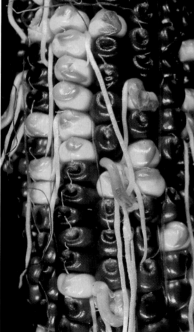

Viviparous germination on an ear of corn.

metabolically destroyed before the buds or twigs can resume growth in the spring. Since the ABA is sometimes destroyed only at precise temperature regimes, its destruction can be easily controlled or mimicked by physical or chemical treatments. Growers sometimes apply such treatments to fruit trees especially to maximize productivity. The timed control of seed germination and bud growth by ABA and other compounds is a widespread mechanism favoring plant survival.

It appears that abscisic acid and gibberellic acid can sometimes act antagonistically. For example, ABA inhibits stem growth and gibberellin promotes it; ABA promotes dormancy and gibberellic acid relieves it; in some long-day and short-day plants, the two hormones have opposite effects on flowering. Since both ABA and the gibberellins are derived from a common chemical precursor, mevalonic acid, the biochemical branch point leading to the synthesis of one hormone or the other determines many aspects of the plant's subsequent behavior.

Ethylene

Ethylene, a gas whose molecules contain only two carbon atoms, is the simplest of the known plant hormones. At the beginning of this century, commercial growers were worried about the yellowing, leaf fall, and premature senescence of plants growing in artificially heated greenhouses. A few simple experiments revealed the cause of the damage: it was induced by a product formed in the incomplete combustion of the natural gas used to warm the greenhouses. The effects were so costly that chemists began an intense exploration to discover which product of incomplete combustion was producing the damage. Their studies culminated in the demonstration that ethylene was the causal agent.

Ethylene

Ethylene is the chemical that signals a green fruit to ripen. Sturdy green bananas are shipped from tropical regions to warehouses in the United States, where they are treated by a puff of ethylene approximately two days before their desired appearance in the market as soft, sweet, yellow fruits. In the 48 hours after their exposure to ethylene, the fruits become sweeter as reserve starch is broken down to sugar, softer as the cell walls are digested, and tastier as aromatic substances form that give the fruits their distinctive flavor.

The ripening ability of ethylene is exploited by rot-inducing fungi that infect plant parts, especially fruits. For example, the fungus *Penicillium digitatum* growing on oranges produces such copious quantities of ethylene that not only the host fruit, but also its neighbors, become rotten. The

The persimmons at the left were stored in the paper bag with the apple slices; the apple released ethylene, and under the influence of the gas the persimmons ripened. The persimmons at the right were not exposed to ethylene and remained green.

fungus then advances easily through the softened plant tissue. Similarly, an aging, uninfected apple that starts to rot spontaneously produces ethylene that can spoil the other apples stored with it. This is the origin of the well-known saying that "one rotten apple in the barrel spoils them all."

A fruit forms from the ovary, the egg-containing female sex organ found in the center of the flower. It is surrounded by the male sex organs, the anthers, which produce the pollen grains that give rise to the sperms. Around the anthers are the fragrant, colorful petals and sepals that attract insects and other animals useful in transferring pollen. When a pollen grain is transferred to the stigma, a sticky receptive surface at the top of the ovary, it germinates, producing a tube that grows down the style into the ovary. Chemically guided, it approaches the sac, or ovule, in which the egg is found, then bursts, liberating a pair of sperms. One sperm fertilizes the egg, to form a cell that divides repeatedly to become the embryo, while the second combines with other nearby nuclei to form a cell that gives rise to the endosperm, in which reserve food is stored. The ovule wall ripens to form a protective seed coat that encloses the embryonic plant and the endosperm.

While all this is happening, auxin is diffusing from the seed. This hormone stimulates the cells in the surrounding ovary wall to divide repeatedly and enlarge, forming a green fruit. When ethylene contacts the green fruit, ripening occurs. The colorful and flavorful ripe fruit attracts animals that eat the flesh and disseminate the seeds. In this way, the plant's progeny are spread.

In many fruits, a peak of respiratory CO_2 output called the climacteric immediately follows the peak of ethylene production that is the crucial metabolic event leading to ripening. While we do not completely understand the internal chemical events leading to onset of the climacteric, it has been observed that the formation of ethylene in tissues can frequently be induced by high concentrations of applied auxin. An increase in auxin concentration may be the triggering event that normally leads to ethylene production and fruit ripening.

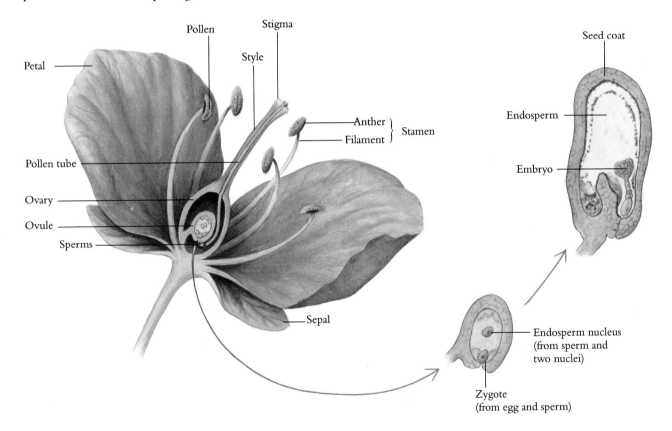

Two sperm travel through the pollen tube and enter the ovule, which contains a sac with eight cells including the egg. One sperm unites with the egg to form a zygote, and the other unites with two cells to form the endosperm. The other cells simply degenerate. The zygote divides repeatedly to form the embryo from which the plant develops. The ovule ripens to form the seed and the ovary ripens to form the fruit.

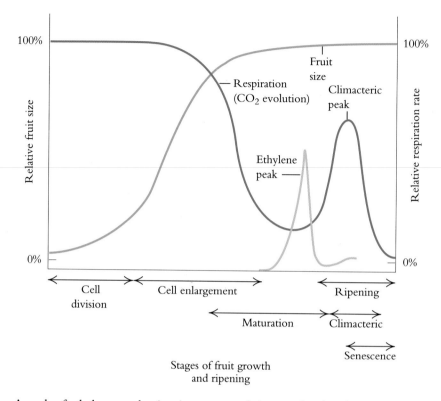

A peak of ethylene production in a mature fruit precedes the climacteric, a burst of carbon dioxide release that signals ripening.

The formation of the fruit from the ovary wall is the only part of the process that an external auxin spray can accomplish. Auxin cannot initiate the maturation of the ovule, and this is the reason that fruits formed by spraying auxin on an unpollinated ovary lack seeds. Genetic lines have been developed that produce a high auxin concentration by themselves; these require neither auxin sprays nor ethylene treatment. If ethylene is required for fruit ripening, it can be supplied either as gas from a cylinder or in the form of a commercial water-soluble chemical such as Ethrel, which is converted to ethylene in plant cells.

Ethylene can occur naturally in plants, or it can be induced to form as a result of a fungal infection, an auxin spray, or even a mechanical injury such as a cut. However it arises in the plant, ethylene in high concentrations can trigger a series of abnormal physiological responses including stunting, stem thickening, premature leaf abscission, loss of chlorophyll, formation of other pigments, failure to respond properly to gravity, pre-

The strawberry "fruit" is actually a swollen stem whose growth is regulated by the release of auxin from the achenes (the seeds, or true fruits) (*top left*). When the achenes are partly removed (*top right*), the fruit does not develop well. When all achenes are removed, the fruit does not develop (*bottom left*); but when such a strawberry is sprayed with auxin (*bottom right*), development is near to normal.

mature flowering, and, of course, fruit ripening in some species. It is probably common for plants grown in enclosed spaces to be damaged by ethylene that has accumulated in the air; excised plant parts, such as fruits, stored in closed containers may be similarly injured. Some of the damage produced by ethylene can be prevented by a high (about 5 percent) concentration of carbon dioxide, and it is now common to store apples in such an atmosphere.

The abscission of leaves that accompanies the approach of winter is usually photoperiodically controlled. Since abscission is stimulated by ethylene, it is assumed that ethylene production can sometimes be controlled by phytochrome. In abscission, ethylene first stimulates the production of a layer of thin-walled cells, the abscission layer, where the petiole meets the stem. The fragile walls of these cells come apart when further weakened by the action of the wall-digesting enzyme cellulase, and the weight of the blade causes the leaf to be shed. This is the only time in the life of

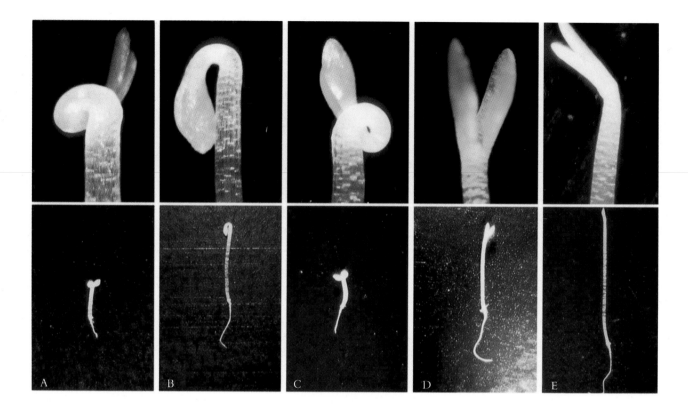

The apical hook of an Arabidopsis seedling (B) becomes severely curved under the influence of ethylene (A). In a mutant (C), severe curving occurs spontaneously, without the application of ethylene. Other mutants (D and E) respond atypically to ethylene by straightening their hooks completely or partially. Thus, genes influence the response of plants to hormones such as ethylene.

the plant when cellulase functions; this enzyme is otherwise unseen in the plant except as a product of invading fungi released to disrupt plant tissue. Curiously, ethylene is frequently formed as a result of the injury associated with such fungal invasions; this "wound" ethylene activates genes controlling the production of compounds harmful to the fungus. Thus, ethylene and cellulase can each, under particular conditions, induce the formation of the other substance.

Ethylene is synthesized in plants from the common amino acid methionine. With the help of appropriate enzymes, methionine first reacts with ATP to form a metabolic product called S-adenosylmethionine (SAM). SAM is in turn broken down to aminocyclopropylcarboxylic acid (ACC), which liberates ethylene directly. SAM also helps to produce a group of chemicals called polyamines, which can inhibit ethylene biosynthesis and are generally antagonistic to its action. The metabolic fate of SAM may thus control whether a fruit ripens or not and whether a plant organ senesces or not. Since some tomato varieties whose fruits have abnormally long shelf life contain greatly elevated polyamine levels, it appears that polyamines may normally antagonize ethylene in the plant. This two-way control of ethylene production increases the plant's ability to fine-tune its effective hormone level. Polyamines are currently finding some use as antiripening agents that increase the shelf life of harvested tomato fruits.

How Do Hormones Act?

Hormones are so dilute in the cell that in some instances they could not form a film even a single molecule thick over the cell's surface. How can such minute quantities of these regulatory substances produce such large effects? Some sort of amplification mechanism is necessary, but the precise mechanism remains a mystery.

We can envision three possible sites for amplification, the same three hypothesized for phytochrome. The most obvious is probably the nucleus (and to a certain extent other DNA-containing organelles such as chloroplasts), which contains the DNA that encodes the various structural and catalytic proteins of the cell. A hormone able to switch genes on or off could control the cell's assembly of structural proteins and enzymes, and thus its gross biochemistry. As we shall see, there is now impressive evidence that several plant hormones can indeed activate specific genes to produce specific proteins. In most instances, however, we do not know the exact function of the proteins produced. Because of the rapidity (sometimes as little as 3 to 5 minutes) with which such proteins are formed after hormone application, we assume that they are somehow connected with the physiological effects produced by the hormone.

A potential second site of amplification is the cytoplasm, in which resides the vast assembly of enzymes that control the gross biochemistry of the cell. In some cases, an "immature," inactive enzyme is transformed into a catalytically active molecule through the attachment of a foreign molecule. If that molecule is a hormone, then amplification is achieved by the simple fact of attachment, for the newly active enzyme magnifies the effect of the single hormone molecule by catalyzing the same reaction over and over. We have convincing evidence that hormones do become attached to specific proteins, but we still do not have unequivocal proof that such attachment confers on the protein a catalytic activity that was not already present.

A third potential site of hormone action is one of the cell's many membranes. For example, a hormone could control events inside the cell by acting on a component of the plasma membrane that regulates the entry or exit of a particular ion, such as calcium, potassium, or hydrogen.

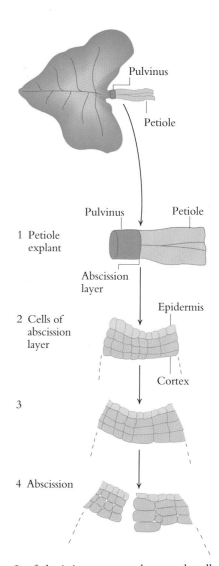

Leaf abscission occurs when weak cell walls at the base of the petiole produce ethylene (2), which triggers the formation of cellulose. The cells enlarge (3) and are partly digested by cellulase (4), causing the fall of the leaf.

Even small changes in the concentration of these ions in the cytoplasm can alter the rate at which important cellular events occur; these events can in turn control many aspects of the fate of cells. Again, there is considerable evidence that this kind of control can be exerted by some of the plant hormones; abscisic acid, for example, seems to regulate the opening and closing of stomata by controlling the passage of potassium through the guard cell membrane.

Auxin is the only hormone whose mechanism of action has been well investigated, so we shall use it as our example. Yet even of auxin's many effects, only its promotion of cell extension is at all well understood.

The roughly cube-shaped young cell typically affected by auxin contains numerous small, scattered vacuoles. These cellular compartments, surrounded by membranes, can gain or lose water osmotically in response to changes in their internal content of salts, sugars, and other dissolved materials. As explained in Chapter 1, sugar or salt molecules within any membrane cause the net inward movement of water molecules by interfering with their free diffusion outward through the membrane. By contrast, pure water outside the cell suffers no such restraint (or at least less

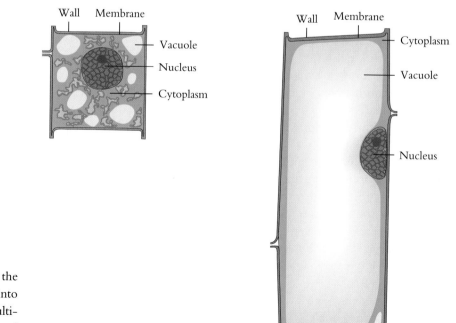

Cubical cells (*left*) produced at the meristem expand by taking water into numerous small vacuoles, which ultimately fuse into a large central vacuole (*right*).

restraint because the concentration of interferring solute molecules is lower) and moves inward at a rate exceeding the rate of outward diffusion. As a result, water accumulates in the cell, and the internal pressure against the cell wall becomes greater.

Treatment with auxin induces a massive uptake of water, which eventually finds its way into the vacuoles. Thus, in the fully enlarged cell, all of the vacuoles will have coalesced into a large central tank constituting 90 percent or more of the total volume of the cell. The increased volume of the cell's contents raises the pressure against the rigid cell wall which stretches slightly; the cell elongates until no more water can enter because of the tremendous back pressures encountered.

The massive uptake of water is made possible mainly by a loosening of the structure of the restraining cell wall, partly through a deposition of new wall material at critical locations. We can envision the water as air and the plant cell as a basketball containing an expanding balloon constrained by a slightly expansible, semirigid jacket. It takes pressure to enlarge the ball, especially after the internal balloon starts pressing against the outer layer, and eventually the back pressure of the jacket prevents more air from entering. The pressurized internal balloon will expand further only if the rigidity of the outer layer is loosened, either by the slippage of the joints holding the material together or by the insertion of new material into the fabric structure.

Osmotically generated pressure enlarges the cell by pressing against, and thus stretching, the cell wall. If a sudden loss of water were to diminish this pressure, part of the wall extension would be automatically reversed, but another part would remain. The amount of the extension that is reversible is the elastic component of deformation; the amount that is irreversible is the plastic component. Early in the pioneering era of auxin studies, Anton Heyn, working at the University of Utrecht in the Netherlands, showed that treating cylinders excised from oat coleoptiles with auxin increased the plastic component of their deformation. Heyn hung the small weight of a balance on one end of a coleoptile cylinder suspended on a pin, so that the cylinder bent down. When the weight was removed, the cylinder returned elastically only part way toward its original position. Cylinders treated with auxin showed a greater irreversible (plastic) component of the bending induced by the weight than did untreated cylinders. This simple observation has been confirmed and quantified in more recent years by the use of industrial strain-testing machines.

Chemical analyses have revealed that small quantities of new carbohydrate-like materials of various kinds are inserted into the cell wall during auxin-induced growth. These molecules usually are not mainly cellulose,

Heyn's experiment, showing that auxin increases the plastic extension of the wall of a coleoptile.

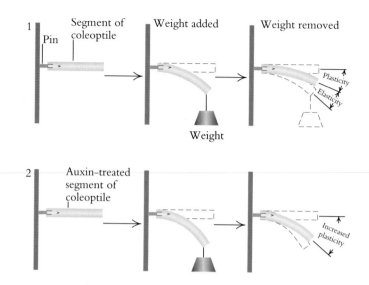

the major structural polymer of the cell wall, but rather hemicelluloses that join the cellulose fibers into an integrated network. By adding small quantities of hemicelluloses at crucial junction points, one could imagine increasing the stretchability of a latticelike structure.

Some researchers have gathered impressive evidence suggesting that during auxin-induced wall stretching, protons (hydrogen ions) are secreted from the inside to the outside of the cell through the action of an ATPase, an enzyme in the membrane that hydrolyzes ATP with the release of energy. The secretion of protons would acidify discrete regions of the cell wall, and this acidic condition would activate or speed up the rate of the enzymatic reaction in which new links are inserted between cellulose fibers.

In addition to activating already existing enzymes, auxin initiates the transcription of messenger RNA at particular genetic loci. The rate of synthesis of certain discrete proteins is considerably altered when auxin is administered to responsive tissue. Some of these proteins become more abundant, others less so. Of the approximately half a dozen proteins that have been demonstrated to rise rapidly in quantity in response to auxin, two appear to be formed so quickly that they could play a role in the growth induced by auxin. Unhappily, we do not yet know what these proteins do, or even whether they are structural proteins or enzymes.

The presence of auxin increases the concentration of calcium in the cytoplasm of some cells, probably by virtue of auxin's attachment to a specific receptor protein located in the plasma membrane. This hormone receptor complex then interacts with another protein, which is capable of binding guanosine triphosphate (GTP), an analog of the high-energy compound ATP. Such "G proteins" are frequently involved in linking external stimuli, like light or hormones, to the chemistry of events inside the cell. The attachment of the auxin-receptor complex to the G protein in the membrane activates a series of events that leads to the formation of inositol trisphosphate (IP_3), already discussed in Chapter 2. This compound opens calcium channels in the membranes of calcium-rich cellular organelles, thereby increasing the calcium content of the cytoplasm. This extra calcium may itself interact with another receptor protein, called calmodulin; the calcium-calmodulin complex initiates another cascade of regulatory events connected with the action of calcium itself. For example, protein kinases, enzymes that control the activity of other enzymes by attaching phosphate groups to them, are often stimulated by calcium. Thus, increasing calcium concentration by these indirect means could control the entire chemistry of the cell. This scheme, based mainly on data from animal and microbial systems, has not yet been demonstrated to occur in plants; nevertheless, many botanical researchers seek to apply it to plant cells in the belief that basic biochemical mechanisms are similar in all creatures.

Our understanding of other hormones is even more limited. As discussed earlier, gibberellin turns on genes controlling the production of amylase and other enzymes. Comparatively little is yet known about the molecular basis for the action of cytokinins, abscisic acid, and ethylene; but in each case, it has been suggested that the hormone molecule binds with a receptor protein as the first step in a cascade of events. In all cases, patterns of protein synthesis and enzymatic activity appear to be altered, probably, as with auxin and gibberellin, because specific genes have been activated.

Rooted in a fixed location, plants may be considered immobile, yet their parts move all the time. Plants move as they grow, as they respond to gravity, light, and other stimuli, as they transport nutrients, and as they carry out pollination and seed dispersal. Focus a microscope on certain plant cells and you will see chloroplasts and other cellular bodies moving about briskly in orderly but changeable channels, sometimes extending the entire length of a cell. Train a time-lapse movie camera on a growing stem tip or root tip and you will observe the constant side-to-side oscillation that constitutes circumnutation. Train the camera on the leaves of beans and other leguminous plants and you will capture a rhythmic opening and closing movement, like the flapping of birds' wings. Place a vigorously growing seedling in a horizontal position and you will quickly note the upward curvature of the stem tip and downward curvature of the root tip. Point a beam of light to one side of a stem tip, and the stem will curve toward the light. All of this movement is accomplished without the nerves, muscles, bones, ligaments, and tendons on which animals depend. All a plant needs is control over patterns of growth and the movement of water and salt, and sometimes the assistance of microscopic fibers of proteins, similar to the characteristic components of muscle.

FOUR

The Movements
of Plants

Cytoplasmic Streaming

Although the plant cell does not contain muscle fibers, it does contain the two major protein components of muscle, actin and myosin, which interact to produce contractions and movement in much the same way as they do in animals. With a few exceptions, however, the large-scale movements of plants observable with the naked eye are performed without their participation. Instead, the interaction of these proteins is most important in certain energy-requiring movements, visible only with a microscope, that occur continually in the interior of plant cells.

In many cells, the cytoplasm is in a constant state of rotation (cyclosis) around the cell wall. The rotating fluid follows pathways that branch, join, and constantly change, but the stream maintains a regularity of movement in a given direction for long periods of time. The movement can easily be studied by observing organelles, especially chloroplasts and mitochondria, that are carried along passively with the moving stream. This passive drifting is to be distinguished from the independent, active movement of chloroplasts like that caused in the alga *Mougeotia* by the transformation of phytochrome.

Not all of the cytoplasm participates in cyclosis: a nonmobile outer (ectoplasmic) layer of the cytoplasm separates the moving inner (endoplasmic) layer from the cell wall. The cell's organelles appear to stream along fibrils about 0.2 μm in diameter located at the boundary between these two cytoplasmic layers. The fibrils are oriented parallel to the direction of streaming and do not themselves move; only the organelles and the colorless stream bathing them are transported.

The fibrils guiding cytoplasmic streaming are part of the cytoskeleton, a network of fibers that gives the cytoplasm some form and rigidity. Each fibril appears to consist of a bundle of microfilaments, each about 6 nm in diameter, composed of strings of the protein actin, one of the two components of animal muscle fiber. Actin is abundant in many plant cells, and its tiny fibers sometimes constitute up to 3 or 5 percent of the total protein of the cell. The other component of muscle, myosin, is also present in the cytoplasm of some plant cells, often in association with actin as part of an actin-myosin complex. It forms thicker fibers and can change form as it breaks down ATP. In this action, myosin acts as an enzyme, called an ATPase, that makes the energy of ATP available. If actin is also present, the two proteins associate closely, causing the ATPase activity of myosin to be greatly enhanced. As the ATP is degraded and myosin changes form, projections of the myosin molecule, resembling gear teeth, push against

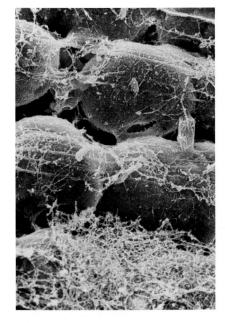

Rows of streaming chloroplasts in a cell of the alga *Nitella* are connected to long actin fibers. Under the influence of ATP and myosin, the chloroplasts are carried around the cell by the actin-myosin interaction.

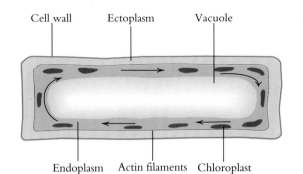

Cell wall   Ectoplasm   Vacuole

Endoplasm   Actin filaments   Chloroplast

The inner, endoplasmic layer of cyto-plasm streams around the central vac-uole (*top*), driven by the action of myosin and actin (*bottom*). Using the energy of ATP, the myosin is thought to move along the actin fiber by re-peatedly binding to the fiber and pulling against it.

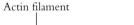

Actin filament

Myosin molecule

grooves in the actin fiber, moving the myosin along the fiber. If myosin molecules line up in an orderly fashion along actin fibers, the sliding myosin propels the entire viscous cytoplasm.

Streaming facilitates the transport of slowly diffusing large molecules and organelles from one end of the cell to the other. The long cells in algae and in the vascular tissue of some stems and roots have the most vigorous cytoplasmic streaming; in these cells, the rate of streaming some-times exceeds 50 μm per second. Even though some plant cells are 1000 μm or more in length, streaming thoroughly mixes the contents of these giant cells within a fraction of a minute.

Circumnutation

We have already drawn attention to the circumnutational movements of plant organs, the ceaseless spiraling of stems, roots, and tendrils around a central axis described by Charles Darwin during his investigations of tropisms. Impressed as he was by the powers of natural selection, Darwin concluded that these oscillations are a necessary antecedent to the tropistic curvatures that are required by a successfully adapted land plant in response to light, gravity, and other stimuli. As we have seen, tropistic curvatures are a modification of circumnutational patterns from a symmetrical spiraling about a central axis to an asymmetrical curvature toward or away from a stimulus.

The cause of circumnutational movements is not known. One theory suggests that they are the result of random inequities in growth: a temporary spurt on one side of the stem's vertical axis leads to a displacement toward the other side. Perception of gravity in the displaced tip then causes a righting reaction and an overshoot, which leads to displacement toward the other direction, and so on. The tip gradually becomes vertical again as the overshoots diminish in amplitude. This theory, however attractive, is probably rendered invalid by the recent observation that plants transported to the low gravity of outer space still show circumnutational movements. An alternative theory holds that the cells of any growing organ elongate according to intrinsic rhythms, and that circumnutation is just a visible manifestation of these rhythms. Fortunately, scientists have made greater progress in understanding the tropistic curvatures toward light and gravity, and it may be that further studies of tropistic movements will eventually illuminate the mechanisms of circumnutation.

Phototropism

The bending of plants toward light seemed for many years to be satisfactorily explained by the Went-Cholodny theory, at least for coleoptiles. The theory holds that light induces an asymmetry in the export of auxin from the coleoptile tip, and that this asymmetry is the cause of the curvature. Went's discovery that the darkened side of a coleoptile exports twice as much auxin as the lighted side, and that this imbalance coincides with or precedes the curvature of this organ, would seem to make a logical case that curvature is controlled by auxin. Circumnutation could result from transient and changing asymmetries in the release of auxin from particular groups of apical cells, and the circumnutating organ could show tropistic

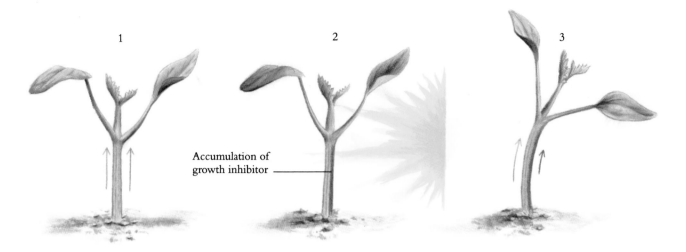

Accumulation of
growth inhibitor ————

Phototropic curvature might be caused by the differential production of a
growth inhibitor under the influence of light. Whereas a stem in a symmetrical light field grows straight up because growth inhibitors are symmetrically
distributed (even shading) (1), asymmetrical light causes a growth inhibitor to
accumulate on the illuminated side (darker shading) (2). Growth on the illuminated side is slowed relative to the shaded side, and the stem curves (3).

curvature when an asymmetric stimulus such as light leads to a prolonged,
stable asymmetry.

   This theory seems to fit coleoptiles quite well, but is less successful
when applied to other organs, such as the stems of sunflower and other
dicotyledonous seedlings, which, unlike coleoptiles, have substantial
green pigmentation. If auxin transported downward from the tip were
simply shunted to the shaded side of the stem, that side should grow faster
than normal. Yet researchers in England, the Netherlands, and Japan have
reported in recent years that although growth on the lighted side of such
green stems is inhibited when the stems are exposed to asymmetric light,
it is not stimulated on the shaded side. The researchers hypothesize that
light causes a substance that inhibits growth to be produced from some
innocuous precursor. The shaded side contains less inhibitor and thus
grows more rapidly than the illuminated side, but no more rapidly than a
symmetrically illuminated control.

   In support of this theory, these researchers point to the existence of
several substances, extractable from these stems, that seem to increase in
concentration rapidly after illumination. Not only are these substances
more abundant on the lighted side of the stem, but in tests of even tiny

quantities, the substances will inhibit the straight growth of cylinders of the same plant material. One such material, xanthoxin, seems to be produced from carotenoid molecules present in the stem. Another, raphanusanine, is present in radish (*Raphanus*) seedlings. Supporters of this theory also point out that the auxin asymmetries proposed by the Went-Cholodny theory either do not appear at all in sunflower, radish, and other dicotyledonous stems or appear only after the stem has already curved. Therefore, they suggest, auxin asymmetries cannot be the cause of curvature.

It appears that those who oppose the Went-Cholodny theory are now in the majority, but more classically trained plant physiologists are not willing to abandon completely the theory, which seems to apply so well to coleoptiles. At the very least, they would retain it for coleoptiles and perhaps other lightly pigmented organs. Even in green organs, they maintain, inhibitors may act by interfering with polar auxin transport. For example, a greater concentration of inhibitor on the lighted side of a stem might more greatly deter the downward movement of auxin on that side; the shaded side would receive more auxin, and its disproportionate growth would cause the stem to curve toward the light. Molecules, both natural and synthetic, that inhibit polar auxin transport are known to exist, including an entire class of substances, called morphactins, that influence the morphology of plants by interfering with auxin transport.

We have seen that light, in order to be effective, must first be absorbed by a pigment. The clear peaks in the red and the blue regions of the action spectrum for photosynthesis implicate chlorophyll as the photoreceptor, while the reversal of the effects on seed germination as the light changes from red to far-red clearly indicates phytochrome as the photoreceptor. Similar experiments with phototropism have revealed that the blue region of the spectrum elicits curvature most effectively. Early action spectra showed a peak in the region of 440 to 480 nm; wavelengths longer than about 500 nm were completely inactive. Subsequent spectra displayed a multiplicity of peaks. Because the match is fairly close with the absorption spectra of carotenoids, and because carotenoids are the effective pigments in vision, many researchers automatically assumed that a carotenoid is the receptor pigment for phototropism.

In the late 1940s, when as a young investigator working at Caltech I began looking into the effects of light on plant growth, I fortuitously discovered some interesting effects of another yellow pigment that absorbs blue light in the same spectral region as carotene and the unknown phototropism receptor. I found that riboflavin (vitamin B_2) could act as the receptor pigment for several light-activated reactions, including the oxidation of indole acetic acid, the amino acid tryptophan, and related com-

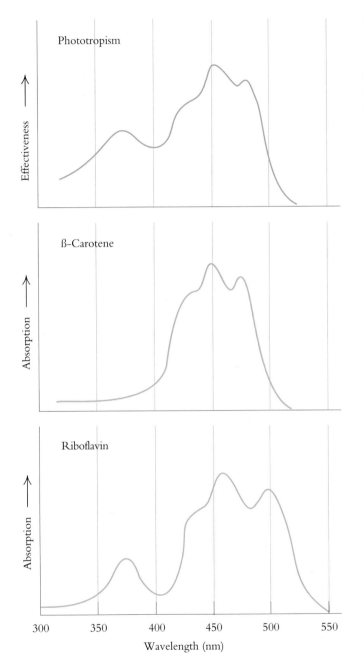

Phototropism

ß-Carotene

Riboflavin

Wavelength (nm)

Both the action spectrum for phototropism (*top*) and the absorption spectrum for riboflavin in oil (*bottom*) show a peak near 370 nm; the absorption for beta-carotene (*middle*) lacks this peak. The other peaks fit carotenoids somewhat better.

pounds. I hypothesized that a similar reaction, the oxidation of auxin, could be involved in phototropism and that riboflavin as well as carotene deserved consideration as a photoreceptor for phototropism. In the years

Curvature, and straightening, occur because of differential growth. Here a cress seedling is straightening its apical hook after illumination. The black dots are resin beads on the outside of the stem; the changing distances between beads shows the underside of the straightening stem growing more than the opposite side.

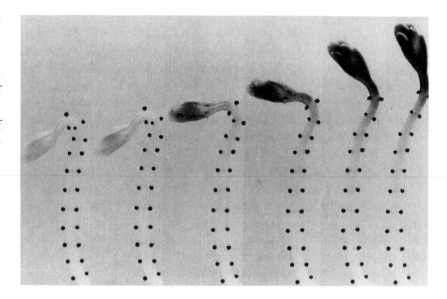

since that first suggestion, the theory of photoreception by riboflavin has found considerable support, although the pigment probably does not act by oxidizing auxin.

Numerous investigators pointed out that when riboflavin couples with proteins to make flavoprotein enzymes, the absorption spectrum is altered somewhat and the new spectrum matches that for phototropism even more closely. Thus, a flavoprotein, rather than free riboflavin, is more likely to be the sought-after pigment. Recently, Winslow Briggs and colleagues at the Carnegie Institution in Stanford have found flavoproteins in various cellular membranes; phosphate groups become attached to these flavoproteins when the enzymes are light-activated. These flavoproteins could be the photoreceptors for phototropism, but more evidence is required to settle the question.

In attempting to evaluate carotenoids and flavin-type compounds as possible photoreceptors, action spectra have of course been very useful. Although both types of pigments match the action spectrum closely in the visible part of the spectrum, from 400 to 480 nm, their absorption spectra diverge in the near ultraviolet. Their absorption in that region of the spectrum supports the candidacy of riboflavin. Action spectra for phototropism show a clear maximum near 370 nm, a peak that also appears in the absorption spectrum of most flavoproteins but is absent from those of the most common naturally occurring carotenoids.

This critical evidence in favor of flavins seemed decisive until recent work with mutants of the plant *Arabidopsis thaliana*. Certain mutants of this plant, called "phototropicless," do not bend toward blue light shone on one side of the stem, but blue light does markedly inhibit the elongation of their stems. Other mutants show just the reverse behavior. Since the two responses are sensitive to slightly different spectral regions, the action spectrum for phototropism may be a composite, reflecting contributions from two different types of photoreceptor pigments. This leaves carotenoids as a continuing possibility, with flavins also likely candidates. However, certain mutants of both higher plants and the fungus *Phycomyces blakesleeanus* still react normally to unilateral blue light, even though they are completely or almost devoid of carotenoids. Unless tiny, undetectable quantities of carotenoids are extraordinarily effective receptors, this evidence strongly argues that phototropism can occur in the absence of carotenoids. It thus favors flavins as the photoreceptor pigments, but does not rule out that carotenoids or combinations of other blue-absorbing pigments may also serve when present.

Gravitropism

When a farmer tosses a seed into the ground, he need not worry about the stems coming up and the roots growing down, for the plant has automatic mechanisms guaranteeing the correct direction of growth. As in phototropism, one side of the cylindrical organ grows faster than the other, curving the tip, and, also as in phototropism, the direction of curvature depends on an asymmetrical stimulus. The stimulus in this case is the gravitational field of the earth, generally designated as 1*g*. When, for example, a grass seedling is laid out in a horizontal plane, the coleoptile tip curves upward, away from the direction of gravity's pull, and the root tip curves downward, in the direction of gravity. Gravity no longer exerts a pull in a single direction, however, if the horizontally placed seedling is rotated about its axis on an instrument called a clinostat. In this case, the gravitational stimulus is effectively annulled, and both organs grow horizontally, in the direction of their placement.

Gravitational curvature is mimicked in plants that are moved rapidly around in a circular pathway. For example, the coleoptile of an erect seedling fixed to the circumference of a rapidly rotating wheel responds by curving outward, while the root curves inward. The rotating plant is subject to a centrifugal field in which particles of different mass move at different speeds; as a consequence, the heavier particles separate from the

Gravitropic curvature of the root and coleoptile of a horizontally placed corn grain (*left*) bring these organs into an erect position, like that of a vertically oriented grain (*right*).

lighter ones. Assuming that the curvature of the spinning plant is somehow induced by this separation of particles based on mass, researchers have searched for a related mechanism to explain gravitropism. Many researchers have suggested that the first event in both instances is the fall of some heavy body, called a statolith, through the cell. If gravity is the stimulus, the statolith falls toward the bottom of the cell, while if centrifugal force is the stimulus, the heavier particles, including the statoliths, will move toward the outside. That both responses have a common mechanism is supported by the fact that when gravitational and centrifugal forces are delivered at right angles to each other, the plant effectively integrates both stimuli into a single response. Consequently, when a $1g$ force is applied downward at the same time as a $1g$ force is applied horizontally, as on a rotating wheel, a coleoptile will curve upward at a 45° angle from the vertical and a root tip will curve downward at the same angle.

Statoliths are generally assumed to be especially heavy amyloplasts containing multiple starch granules. In the stem, these bodies are found in cells surrounding vascular bundles; in the root, they are found in the cells

When a cell containing starch-filled amyloplasts is tipped on its side, the amyloplasts slide to the new "bottom" of the cell.

1 VERTICAL ROOT

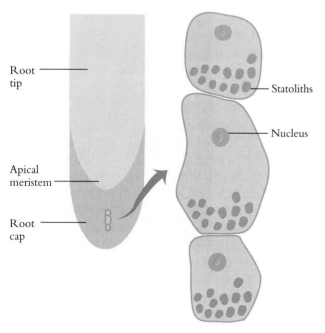

Root tip

Apical meristem

Root cap

Statoliths

Nucleus

Three root cap cells

2 ROOT TURNED ON ITS SIDE

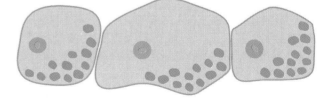

of the root cap, a group of cells at the very tip of the root, which are continuously produced by the root meristem and continuously sloughed off as the root makes its way through the soil. When the plant is inverted or turned on its side, the statoliths fall to the new "bottom" of the cell, and their displacement is soon followed by gravitropic curvature. Yet, starch grains are obviously not the only statoliths capable of perceiving gravity. In the single-celled "roots" of an unusual alga, *Chara,* which lacks amyloplasts, crystals of barium sulfate can apparently perform the same function. In various plants, mutants lacking multiple starch grains in their amyloplasts generally fail to respond well to gravitational stimuli, but the few exceptions suggest that while heavy starch-containing amyloplasts are sufficient to induce gravitropic curvature, they are not necessary. Other organelles of high density may act as statoliths in their place.

Whatever the nature of the statolith, how does its displacement in the cell lead to the observed asymmetry of growth and curvature? As was the case with phototropism, the elongated cells that cause the stem to curve are several millimeters away from the statoliths; similarly, the center of curvature in the root is 2 to 3 mm above the root cap. A message to curve must somehow travel from the statoliths to the elongating cells, and, not surprisingly, a hormone seems like the obvious solution.

Auxin is probably the messenger in stems and coleoptiles, where it plays the same role as it does in phototropism. Auxin accumulates on the lower side of stems laid horizontally; that side grows faster and the stem curves upward.

The role of auxin in roots is more problematical. When a root is gravistimulated, the growth of its cells is actually retarded. The root curves because the growth of cells on the lower side of the root is more retarded than the growth of cells on the upper side. Since roots are much more sensitive to auxin than are stems, concentrations that promote stem growth inhibit the growth of roots, and it was proposed that the same displacement of auxin export would thus inhibit root growth rather than promote it. Although auxin does accumulate on the lower side of gravitropically stimulated roots, it does not always accumulate sufficiently or rapidly enough to support this interpretation well.

An alternative theory proposes that root tropisms are produced by asymmetries in the distribution of an inhibitory substance synthesized in the root tip. Some suggest that this inhibitor might be abscisic acid or a related molecule. This theory is supported by the behavior of the roots of certain varieties of corn; when normally oriented, the tips of these roots symmetrically export a growth inhibitor to the cells behind the tips. Such export is displaced to the lower side when the root is gravistimulated,

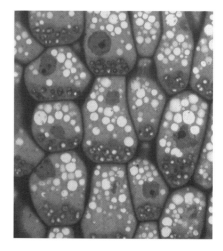

Dense amyloplasts sink to the bottom of root cap cells while lighter vacuoles and oil vesicles rise to the top.

An erect pine twig forms symmetrical patterns of woody cells (*left*), whereas an inclined twig produces reaction wood on the lower side (*right*) that tends to curve the twig upward.

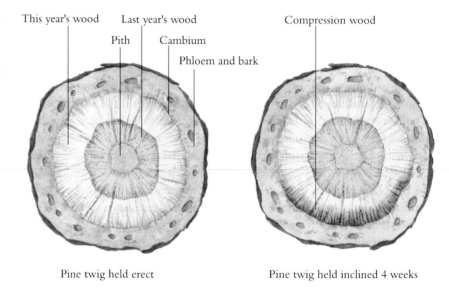

This year's wood · Last year's wood · Compression wood
Pith · Cambium
Phloem and bark

Pine twig held erect · Pine twig held inclined 4 weeks

In broad-leaved trees like maple and oak, an analogous alteration on the upper side of the ring pulls the stem upward, creating "tension wood." The physiology of this bizarre response is obscure, although it may be related to an asymmetric distribution of auxin. Because logs cut from reaction wood contain areas that are under pressure or tension, they tend to warp, and are unsuitable for construction purposes.

Response to Touch and Mechanical Stimuli

As the plant stem grows upward toward the light and away from the surface of the earth, it increases in mass. This new mass added at the top makes the plant top-heavy, and the stem might not remain erect without additional structural support. A growing plant faced with this challenge produces new woody tissues near the base of the stem. The dividing cambium produces tracheid cells, components of the xylem whose cell walls are fortified by lignin, as well as highly specialized support cells called fibers that add considerable mechanical strength to the stem or root. The heavily lignified nature of many plant cell walls has led to the generalization that "plant cells live in wooden boxes." In fact, the thickness of the stem and the amount of lignin formed toward its base are roughly proportional to the mechanical load on the base of the stem, as represented by its mass.

Some plants gain additional physical support without expending the energy needed to build lignified cells. These plants climb upward by fastening "grappling hooks" firmly onto nearby rigid objects such as trees; they save energy by using other organisms as supporting structures. The usual grappling hooks are tendrils—long, slender organs derived from stems, leaves, or flower parts. The typical tendril is approximately cylindrical, with a slight taper toward its apex, which is usually a curved tip.

Tendrils circumnutate vigorously; this movement sweeps them about in circular or elliptical paths, greatly increasing the possibility of their contact with a support. Should the tendril happen to contact a rigid object, it responds within a few seconds or minutes, depending on the species, by forming a spiral coil that will wrap firmly around the object touched. If the touched object is withdrawn, the tendril will uncoil and straighten, but if the touched object remains in place, then the coil continues to develop into a firm anchor for the plant. It is interesting that stimuli such as rain or hail do not eliciting coiling. Although we do not understand why these stimuli are ineffective, it would obviously be immensely wasteful of energy for the plant to respond to stimuli that could do it no good.

Tendrils generally do not attain high sensitivity to touch until they have reached almost their full length, when their circumnutational activ-

The tendril of a vine coils vigorously around the stem of a nearby plant.

ity is also at its highest. Unstimulated tendrils elongate mainly near or at their base, especially when young, but stimulated tendrils elongate mostly at the tip. Thus, the main locus of elongation is transferred from base to tip after mechanical stimulation. The ability of the tendril to curve is greatest near the tip, because as the tendril curves, it also elongates, and the extra growth allows the tendril to envelop the touched object more easily.

Some tendrils will curve in any direction dictated by the location of the touch stimulus; such tendrils are properly called *thigmotropic*. By contrast most tendrils, especially those with a terminal hook, will curve in only one direction irrespective of the location of the stimulus; these are more properly described as *thigmonastic*. (Nastic movements are defined as those whose direction is independent of the direction of the stimulus, in contrast to tropistic movements, whose direction is determined by the location of the stimulus.) If the mature tendril is never touched, it eventually becomes senescent: the tendril develops a loose spiral coil near the tip, sometimes becomes woody, dries up, and falls off.

Tendrils may be completely unbranched, singly branched, or multiply branched; the various types commonly appear at different locations on the same plant. When tendrils are branched, the circumnutational revolutions are generally in the same direction, but in some plants the two members of a tendril pair will revolve in opposite directions. A tendril coiling around a support (contact coiling) usually continues to coil in a single direction. A tendril may also form helical coils along its axis, not necessarily as a result of any mechanical stimulation (free coiling), and a free-coiling tendril frequently reverses the direction of coiling. It is remarkable that the connections between tendrils and their supports will bear very great weights, up to almost a kilogram in the passion flower, although only several milligrams of weight suffice to initiate curving.

In the pea plant, tendrils can be detected coiling within two minutes after contact. A tendril continues coiling for about 30 minutes, then partly uncoils. It remains relaxed for about another 30 minutes, after which coiling can again resume. Changes in length during coiling can be easily ascertained by marking tendrils with India ink before stimulation: the ventral (lower) surface contracts first, and shortly thereafter the dorsal (upper) surface expands. The two surfaces then grow at different rates; the dorsal surface expands rapidly while the ventral surface expands more slowly. Still later, after the tendril has coiled around the support, it strengthens its contact with the support surface by depositing ligninlike materials and developing resin-secreting adhesive pads near the tips. After

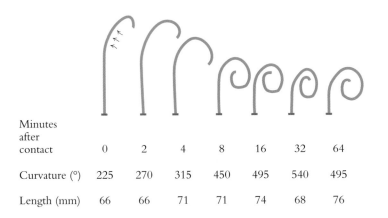

Minutes after contact	0	2	4	8	16	32	64
Curvature (°)	225	270	315	450	495	540	495
Length (mm)	66	66	71	71	74	68	76

A pea tendril curves rapidly after being stroked gently three times (at arrows) with a glass rod.

coiling, the ventral surface usually becomes enlarged, affording greater surface contact between the tendril and its support.

While excised tendrils will elongate in response to both auxins and gibberellins, there is no really good evidence linking these hormones to the coiling reaction. Indeed, the initial stages of coiling are produced not by the differential rates of cell elongation that cause tropisms, but by rapid changes in turgor. Immediately after a tendril is stimulated, it swells rapidly at particular locations and shrinks at others in response to rapid fluxes of water. These fluxes have been studied by observing the movement of tritiated water, which is water that has been labeled by substituting a radioactive isotope of hydrogen (^{3}H) for ordinary hydrogen (^{1}H). Excised tendrils are allowed to absorb tritiated water through their cut base; when mechanically stimulated, they release more of the labeled water from their ventral side than from their dorsal side. This suggests that during the early phases of coiling, when the ventral side contracts and the dorsal side elongates, water leaves the ventral cells and moves downward through the ventral vascular bundle. Any additional water taken up on the rapidly elongating dorsal side of the tendril must have been transported upward through the dorsal vascular bundle.

It is reasonable to expect that changes in membranes may be responsible for the rapid fluxes of water; the membranes could control the osmosis of water by passing or blocking ions. This view is supported by the observation that the efflux of electrolytes (ions that conduct an electric current) is greater from coiling tendrils than from resting ones. The increased efflux is due mainly to an excretion of hydrogen ions (protons), which might be produced by the breakdown of ATP during an ATP-mediated reaction. Contact coiling clearly requires energy, for coiling is greatly diminished when the tendril is cut off from its source of energy by being placed in the dark or the cold, or in an anaerobic atmosphere where ATP cannot be formed. Coiling also slows when tendrils are treated with compounds that inhibit the formation of ATP by photosynthesis or respiration. Whereas tendrils maintained in the dark show a low content of ATP, when they are placed in the light, their ATP content increases, as does their ability to coil, especially if the light is the high-intensity red or blue light typically absorbed by chlorophyll. If excised tendrils are stimulated in the dark by shaking in a petri dish, their ability to coil is greatly increased by adding relatively high concentrations of ATP. Since the coiling of tendrils diminishes the quantity of ATP present, it is reasoned that the coiling response is powered by the ATP produced in photosynthesis or respiration, and that ATP breakdown is initiated by the mechanical stimulus.

How the touch of a tendril on an object would stimulate the breakdown of ATP is unclear, but a role may be played by certain compounds called flavonoids, which are extremely abundant in tendrils. One particular flavonoid decreases dramatically after contact coiling, sometimes declining to about one-third of its former value within an hour. ATP breakdown and coiling come to a halt when this compound is added to a medium in which excised tendrils are immersed. This evidence suggests that the flavonoid may inhibit the breakdown of ATP; when contact between a tendril and a surface causes the destruction of flavonoids, the inhibition disappears and ATP breakdown can provide the energy needed for coiling.

In sum, the destruction of flavonoids after contact would allow ATP to provide energy, either for the movement of hydrogen ions involved in water fluxes or for some other, still unknown mechanism of coiling, perhaps involving actin and myosin. Both actin and myosin have been found in tendrils, and that they act during coiling is suggested by a simple experiment. Tendrils are ground up to form a liquid mush to which ATP and magnesium ions are added. After their addition, the mixture becomes more viscous, a sign that actin and myosin are interacting and using the energy supplied by ATP. The evidence indicates that some sort of contractile process is involved in the curvature reaction. If an actin filament were anchored to the cell perimeter, then a myosin–actin interaction powered by ATP might potentiate coiling by causing a contraction within the cell.

Of all plant organs, tendrils are probably the most reactive and most versatile in their movements, but as we shall see, they are neither the most rapid, nor the most bizarre.

Sleep Movements

As day passes into night, some leaves, especially the compound leaves of tropical leguminous plants, change markedly in appearance. During the day, the paired leaflets are hinged open, presenting a flat canopy that faces the sun, while at night the leaflets fold together, so that very little leaf-blade area is exposed to the sky. What could be the use of such *nyctinastic* movements? Darwin reasoned that this adaption of tropical plants was designed to minimize the heat lost by radiation to the open sky. On a clear night the sky has a very low temperature, approaching absolute zero ($-273\ °C$). Since the heat lost by radiation to such an open sky is proportional to the fourth power of the temperature difference between the leaf and the sky, tremendous cooling can result. Such cooling could be injurious, especially to tropical plants, whose enzymes are regulated to work best at elevated temperatures. The folding together of leaves would minimize heat loss by reducing the radiative surface.

The distinguished German plant physiologist Erwin Bünning hypothesized a different function for nyctinastic movements. The phytochrome system of leaves is sensitive to energy levels similar to those found in bright moonlight, and there have been reports of flowering and seed germination being stimulated in bright moonlight. Bünning reasoned, therefore, that nyctinastic leaf folding represented the plant's attempt to prevent the moonlight from interrupting photoperiodic control. There are as yet no critical data that inform us whether Darwin or Bünning is more nearly correct, or whether some other theory, as yet unstated, will eventually find acceptance.

Sleep movements have been studied intensively in two tropical leguminous trees—*Albizzia julibrissin* (the silk tree) and *Samanea saman.* (When we studied these trees in my laboratory about two decades ago, we referred to them as Al and Sam.) In *Albizzia,* the leaflet pairs fold their ventral surfaces together at night, while in *Samanea,* the leaflets move in the opposite direction and press their dorsal surfaces together. A common mechanism underlies both types of movement.

In both species a fleshy organ, the pulvinus, lies at the base of the petiole or at the bases of the analogous smaller stalks supporting leaflets of the compound leaf. When particular cells are turgid on one side of the pulvinus and flaccid on the opposite side, the petiole will move toward the flaccid side; when the relative turgor of these cells is reversed, the petiole will move in the opposite direction. Using sensitive techniques for measuring ions, such as atomic absorption spectroscopy, it has been conclusively demonstrated that motor cells on the ventral surface of an *Albiz-*

The open compound leaf of *Albizzia julibrissin (left)* closes when leaflet pairs bring their upper surfaces together *(right)*.

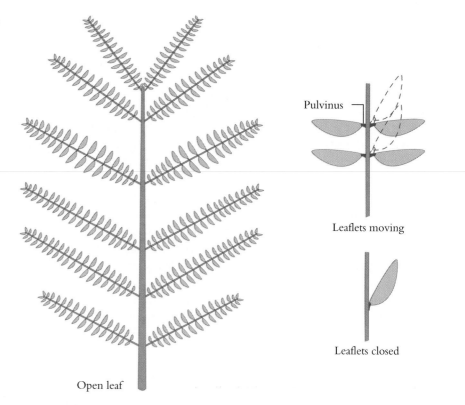

Pulvinus

Leaflets moving

Leaflets closed

Open leaf

zia pulvinus take up water to become turgid after absorbing large quantities of potassium and chloride ions from surrounding cells, and that they give up water to become flaccid when they lose such ions. Thus, there is a push-pull mechanism sending ions flowing back and forth: when ventral cells of *Albizzia* change from turgid to flaccid, causing the leaflet to move upward, the ions they lose will eventually move to the dorsal cells, which then change from flaccid to turgid. The leaflet will move downward again only when the ions have made the return trip and restored the turgor of the ventral cells.

In general, the turgor patterns of the motor cells of the pulvinus are controlled by light-dark transitions, such as those at dawn and dusk. In *Albizzia,* light from the red region of the spectrum promotes opening, while far-red light promotes closure. Thus phytochrome is probably the receptor pigment. At daybreak, the red light that predominates over far-red in sunlight converts phytochrome from the P_r to the P_{fr} form; potas-

OPEN   CLOSED

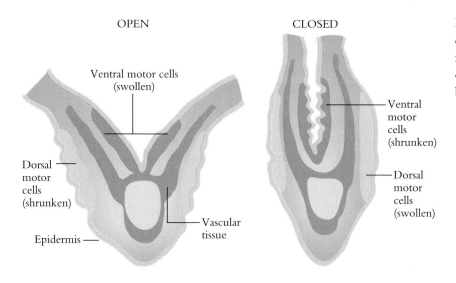

Ventral motor cells
(swollen)

Ventral
motor
cells
(shrunken)

Dorsal
motor
cells
(shrunken)

Dorsal
motor
cells
(swollen)

Vascular
tissue

Epidermis

Leaflets of *Albizzia* open (*left*) and close (*right*) in response to the changing turgor pressure of motor cells on opposite sides of the pulvinus at the base of each petiole.

sium and chloride ions flow rapidly into ventral motor cells that will open the leaflets of *Albizzia*. The far-red light that predominates just as the sun sets converts P_{fr} to P_r, reversing the changes induced by red light and closing the leaflets.

Some chain of events must lead from the activation of phytochrome to the movement of ions into and out of motor cells. Once again, this chain may include the cascade initiated by the liberation of IP_3 and DAG discussed in Chapter 2. This cascade explains both changes in membrane properties and the amplification of the original signal. While the participation of this cascade is still speculative, many features of the suggested biochemistry have been shown to occur in nyctinastic leaves.

Phytochrome alone, however, does not determine the movement of nyctinastic leaves; rhythmic processes also at play may be even more dominant. That innate rhythms also control nyctinastic movements can be easily demonstrated by removing a leaflet pair from an *Albizzia* tree that has been exposed to daily light-dark transitions and placing it in darkness. Under such "free run" conditions, the leaflets will continue to open and close for several days with a period of approximately, but not exactly, 24 hours. This rhythm, approximating one day in length, is circadian, like the rhythms that in part govern flowering.

As is characteristic of circadian rhythms, exposing the plant to light or darkness ahead of schedule resets the rhythm, so that a new cycle of opening and closing begins immediately. A leaf closing in darkness can be

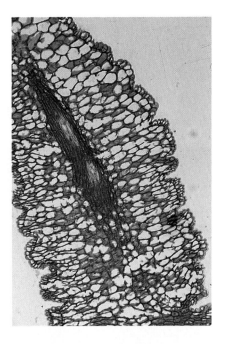

A section through an *Albizzia* pulvinus. Note the swollen motor cells at right, the shrunken motor cells at left, and the central vascular strand.

made to reopen by the properly timed administration of light, red light in particular. Conversely, a leaf opening in the light can be made to slow down or even to start closing by administering far-red light or by transferring the leaf to darkness. The resetting mechanism allows the circadian rhythm to adjust to seasonal changes: the rhythm is reset slightly each day as the time of sunrise and sunset advances or recedes, so that the rhythm remains synchronized with the light-dark transitions. In the resetting reaction, it appears that the photoreceptor for the response mechanism is in the pulvinus itself. This remarkable tiny organ thus contains the photoreceptor pigment, the ions, the water, and all of the biochemical machinery required for the change in turgor that creates movement. For this reason the pulvinus has been employed frequently in investigations of the mechanism of response to phytochrome.

The much studied "sensitive plant," *Mimosa pudica,* exhibits a unique combination of sleep movements and rapid thigmonasty. This plant is virtually indistinguishable from *Albizzia julibrissin* in appearance, but shows a marked difference in its response to touch or a sharp blow: in *Albizzia* nothing obvious happens, but in *Mimosa* every leaf seems to collapse. The sudden limpness is the result of a rapid loss of turgor by key motor cells in the pulvini at the bases of petioles and the small stalks suspending the leaflets.

Mimosa pudica, shown open (*left*), collapses its leaves when mechanically stimulated (*right*).

This striking phenomenon has never been satisfactorily explained but may be the result of a traveling electrical signal called an action potential. When an impermeable membrane lies between two similar salt solutions of different concentrations, a voltage can be measured between the two solutions. The voltage exists because the ions in solution carry electrical charges; the more concentrated solution can carry more charges, creating the charge difference that produces the voltage. A cell bathed in a dilute solution may have a salt concentration inside the vacuole that is 10 times the concentration of the external solution. In this case, the voltage across the cytoplasmic membrane is 58 millivolts (mV). If the concentration ratio is 100 to 1, the voltage rises to 116 mV, following a logarithmic progression. Such transmembrane voltages, or potentials, can be measured by inserting a delicate, needlelike glass electrode into the cytoplasm or vacuole without causing leakage where the needle traverses the membranes. Now, keeping another electrode outside the cell and connecting both electrodes to a sensitive meter, one can observe the voltage. Most plant cells have transmembrane potentials in the range from about 20 to about 100 mV.

Imagine a chain of cells in a row, each with a measured potential of 100 mV; now imagine an excitatory substance hitting one cell and making its membrane leakier. This sudden leakiness will cause an outward flow of ions. As the salt concentrations inside and outside the cell become more equal, the potential of that cell will rapidly lower, and the cell is said to be depolarized. The excitatory substance then moves on, affecting each cell in the line in turn. Immediately after the substance has gone, each cell starts to "repair the damage," and after some time the original potential is restored. The observed sudden change of potential, which moves wavelike down the tissue as the substance hits cell after cell, is called the action potential. Action potentials have been recorded during many types of plant movement. Because ion leakage is usually followed by water loss and decreased turgor, action potentials have often been invoked to explain the movements themselves.

It appears that a blow to the *Mimosa* plant stimulates the production of some excitatory substance that travels rapidly from one location to another, presumably in xylem, generating electrical action potentials as it moves and exciting the pulvini all along the plant to react. Thus, striking one leaf sharply may cause all leaves to collapse. It has been humorously suggested that such a rapid and dramatic change would deter a goat or other herbivore from browsing on *Mimosa*. We know of no other adaptive reason for this bizarre behavior.

Most plants are exposed to potentially harmful stresses every day of their lives. For example, on almost any sunny day, the overheating of leaves poses a threat to photosynthesis. Since the leaf can only cool itself by secreting water through its stomata, any plant undergoing even partial dehydration runs the risk of being cooked in less than an hour. Similarly, toxic substances are a constant problem, whether absorbed from the air by leaves or from soil by roots. The bites of an insect or a grazing animal can cause damaging or even lethal injury, while other, nearby plants can steal necessary water, sunlight, or ground space. Seasonal stresses like freezing temperatures and prolonged darkness affect plants at high latitudes, and random stresses such as flooding, hail, wind, and lightning can threaten the survival of any plant at any time. For a rooted green plant, it is a dangerous world.

Successful wild plants have adapted to these dangers by evolving special structural, chemical, or behavioral modifications that protect them under adverse conditions. A species that fails to evolve an adequate protection against even one stress can face extinction, so the new countermeasures have to be applied quickly and effectively. Responses to extremes of temperatures and to attacks by predators provide especially rich examples of the variety of tactics that have been evolved by beleaguered plants.

FIVE

Coping
with Stress

Staying Cool

Each morning as the sun rises, the plant opens its stomata, carbon dioxide enters, and active photosynthesis begins. At the same time, water evaporates from the surfaces of wet leaf cells and is lost through the open pores. The evaporation of water, called transpiration, is cooling to the plant because some of the leaf's heat energy is used up in converting liquid water to the vapor form. Transpiration usually cools leaves about 3 to 5 °C below the ambient air temperature on a sunny day; in extremely hot climates such as Death Valley, the leaves may cool 8 °C or even more below air temperature.

This cooling protects the leaf from reaching temperatures that would seriously damage its cells. High temperatures are deleterious to organisms because heat energy changes the form of proteins, denaturing them. Since cellular proteins serve either as structural units or as enzymes, their disruption causes major upsets in the cell's structure and metabolism.

Different plants function well in vastly different temperature ranges. Our familiar mid-latitude plants (mesophytes) can grow at temperatures from just above freezing to about 40 °C, the heat of a summer day; thermophiles such as desert plants can extend this range about 10 °C upward; and cryophiles such as arctic plants about 10 °C downward. The enzymes of thermophiles denature at higher temperatures than the enzymes of mesophytes, probably because these plants synthesize special molecules that combine with proteins to protect them against denaturation. In certain thermophilic bacteria, found in hot springs, this protective function is fulfilled by a class of molecules called polyamines. One unusual polyamine, called thermospermine, contains a relatively long chain with many amino (NH_2) groups that form weak bonds with the protein so that its

Stomata close and open rhythmically during successive dark and light periods. The loss of water vapor through the stomata closely follows the pattern of opening and closing, but water uptake lags behind. The stomata of some species close at noon (dashed line) if high temperatures cause transpiration to exceed water uptake.

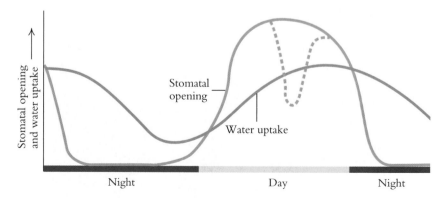

Stomatal opening and water uptake

Stomatal opening

Water uptake

Night   Day   Night

shape changes to a form less vulnerable to denaturation. The existence of similar protective, long-chain polyamines in higher plants is suspected but has not yet been demonstrated. However, in cereal plants threatened by desiccation, salinity, or certain pollutants, smaller polyamines rise rapidly in concentration, and may play a similar protective role.

How much a plant is harmed by higher temperatures depends on the temperature (the degrees of heating above the optimal for the plant) and the duration of exposure. Thus, for many mesophytes, an exposure to 42 °C for one hour can be lethal, while at 50 °C those same plants might die after an exposure of only a few minutes. Whether the damage is lethal is determined in part by physiological factors such as the state of hydration. A dry seed can withstand temperatures in excess of 100 °C for extended periods, while a highly hydrated leaf would be well "cooked" by only a few minutes' exposure to a temperature of 50 °C. This sensitivity to the presence of water parallels the behavior of many proteins, which are stable to heat when dry, but labile when hydrated.

As long as a plant has an adequate water supply, the evaporation of water through the open stomata provides the slight cooling that helps it withstand the warming of the sun. But once the supply of water becomes inadequate, the plant is in danger of dehydrating and overheating. It must now respond to a new stress.

The concentration of the hormone abscisic acid rises dramatically in response to water deprivation. As this hormone accumulates within the guard cells of the stomata, the cells respond by activating a pump in their plasma membrane that excretes potassium and other ions into the surrounding leaf cells. Water follows the ions out of the guard cells in response to osmotic forces, and the pressure of the guard cells' contents against their cell walls diminishes. As the cells lose turgor, the stomata close, sealing the leaf against further transpiration. Thus, the rise in ABA concentration is a safety valve that prevents excessive leaf desiccation. When water becomes more plentiful and the leaf is once again fully hydrated, the level of ABA declines to the unstressed level. The outward potassium pump is turned off, potassium content rises, water enters the guard cells, and the stomata reopen. The importance of this mechanism is seen in certain "wilty" plant mutants. In the tomato mutant *flacca,* which cannot form enough stress ABA, the stomata remain open in dry conditions, and even slight dehydration results in severe wilting. Spraying the plants with ABA restores the normal closure of stomata and protects the plant against desiccation.

The closing of the stomata safeguards the plant against excessive water loss, but may do harm in bright sunlight by depriving the plant of the

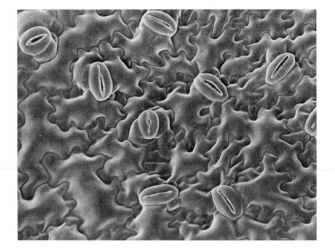
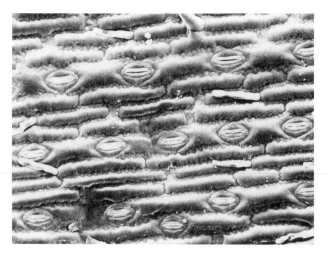

The irregular arrangement of stomata in the leaf of a potato plant (a dicot, *left*) contrasts with the orderly arrangement of stomata in a leaf of corn (a monocot, *right*). The size and placement of stomata tend to optimize their ability to transport gases without mutual interference.

cooling effect of transpiration. Once the stomata have closed, the leaf may attain a dangerous temperature within minutes. It is noteworthy that crassulacean plants, which keep their stomata closed during the day, are nevertheless temperature hardy, able to tolerate temperatures up to about 60 °C. Rather than losing heat through transpiration, these plants apparently reradiate heat directly with exceptional efficiency, and they lose heat efficiently through convection and conduction as well. We do not know what unique mechanism is responsible for this behavior.

The glassed-in spaces of greenhouses create high temperatures and high humidity; the moisture in the air significantly retards the evaporation of water from leaves. Consequently, the ability of transpiration to cool the leaves diminishes. Horticulturists are aware that the high temperatures and humidity in greenhouses can cause heat injury even when the stomata remain open. Clearly, any special adaptations that maximize the cooling effect of transpiration under difficult conditions will help the plant to survive the intense solar radiation. For example, one might imagine that transpiration would be increased by the addition of stomata, yet when stomata lie too close together, the evaporating water creates a zone of exceptionally high humidity around each stoma, and water evaporation into the still air from neighboring stomata may be retarded. Plants that are able to cool themselves efficiently have evolved an optimal shape, density, and distribution of stomata on both the lower and upper epidermis of leaves. There is a great variability among plants in the size, shape, number, and location of stomata, as well as in the size of the stomatal pore during partial closure.

Heat-shock Proteins

When most temperate plants are exposed for more than a few minutes to temperatures above about 40 °C, the synthesis of many normal proteins is slowed dramatically. At the same time, new proteins, called heat-shock proteins, are synthesized in significant quantities. These proteins vary greatly in size, shape, and electric charge and accordingly can be easily separated and visualized by the technique of gel electrophoresis.

This technique depends on the fact that all proteins carry an electric charge and will therefore move in an electric field. A mixture of proteins is inserted into a hole or slot at one end of a thin layer of gel lying on a glass plate. When a voltage is applied between opposite ends of the gel, proteins move through the porous gel toward an electrode. Their relative mobility is roughly proportional to their charge, size, and shape: in general, smaller molecules move through the pores more rapidly, and larger molecules more sluggishly. After some minutes, the differential rate of movement of the proteins has separated them well on the gel, and the electrodes are disconnected. The position of all protein bands is made visible by an appropriate stain, and the molecular size of each protein can be estimated by the distance it has moved from the origin. In addition, the width and density of the stained band can be used to estimate the quantity of each protein.

Studies with gel electrophoresis have established a link between heat resistance and heat-shock proteins. In some cases at least, plants with greater resistance to heat have more of the newly formed proteins, and indeed the quantity of some of the proteins and the degree of heat resistance gained by the plant are well correlated. These heat-shock proteins begin forming within an extraordinarily brief space of time after the critical temperature is reached, frequently within 10 to 12 minutes, and the messenger RNAs for these proteins can sometimes appear within 3 to 5 minutes. We do not understand the mechanism through which the temperature shock can so quickly "turn on" specific genes leading to the production of these proteins.

Heat-shock proteins have been detected in a great variety of organisms, including animals, plants, and microorganisms. In some instances the

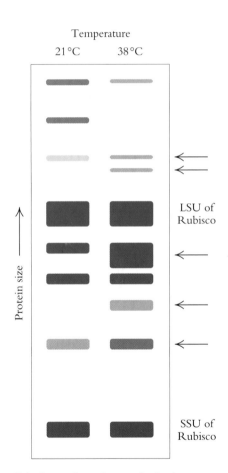

Gel electrophoresis reveals the heat-shock proteins, indicated by arrows, produced more abundantly in barley leaves exposed to high temperature. The large and small subunits of Rubisco, by contrast, are not affected.

under different light conditions: in bright sunlight, the leaves are thicker, but have a smaller surface area for light absorption, while in the shade, they are thinner, with a larger surface area. Thus, by a mixture of structural, chemical, and behavioral tricks, the plant can protect itself, at least in part, from the deleterious effects of an overly hot sun.

Freezing Injury

At subfreezing temperatures, ice crystals can form inside the plant; if their growth ruptures membranes or other vital cellular structures, the affected cell dies, but if membranes are not ruptured, the cell may escape injury altogether. Ice crystals must start to form around a nucleation center such as a particular protein or a cellular particle. In the absence of such a nucleation center, water in the plant cell can supercool, remaining liquid at temperatures below 0 °C. At about −3 to −5 °C, ice usually crystallizes in the spaces between cells, where, in hardy plants at least, the physical displacement caused by crystal growth does little damage to tissues. As long as the cellular contents remain unfrozen, the plant can survive.

The surfaces of many leaves are colonized by bacteria of the genus *Pseudomonas,* which contain a protein serving as a nucleation center. Surface ice crystals, or frost, may form around these proteins and damage tender leaves and buds. In the absence of such nucleation centers on their leaf surfaces, plants start freezing only at somewhat lower temperatures. For tender plants such as the strawberry, the few degrees difference in freezing temperature spell the difference between life and death. For this reason, genetically engineered "ice-minus" bacteria, which lack the nucleation protein, have been produced and patented. It has been proposed that these bacteria be sprayed over strawberry plants in such numbers as to displace the natural bacterial population containing the nucleation protein.

The commercial exploitation of ice-minus bacteria has been delayed by lawsuits filed by environmentalist consumer groups who fear the possible ecological consequences of spreading engineered bacteria in nature. These groups reason that (1) the engineered bacteria might escape into wild populations of plants, possibly making hardy weeds even more difficult to control, and (2) the introduced gene may have unsuspected effects that could produce an agricultural catastrophe. These groups point to the famous case of the Texas "cytoplasmic male sterility (CMS)" gene, which in the 1960s was introduced into corn to facilitate breeding operations. Since this gene causes pollen to be sterile, it abolishes the need to emasculate corn flowers by hand before controlled transfer of pollen from other sources. Unbeknownst to the investigators, however, the CMS gene also makes the plants abnormally susceptible to the Southern corn leaf blight disease. In 1970, a major epidemic of this disease destroyed much of the corn crop, and growers rapidly switched away from CMS-containing corn seed.

One can, of course, never be certain that such "pleiotropic" effects will not appear from any newly introduced engineered gene, but historically such effects have been rare. In any event, small-scale tests should reveal dangerous effects and forestall any major catastrophe. As for the argument that the newly engineered bacteria might spread uncontrolledly in nature, the experimental evidence is convincingly on the other side. Engineered organisms are generally less competent to survive in nature than the wild types of organisms, and generally die out unless repeatedly reintroduced. Since natural bacteria are constantly evolving new variants anyhow, it seems imprudent to place such importance on one man-made new strain differing from the wild type by only one deleted gene.

Chemical Warfare

The world teems with microorganisms, insects, rodents, and other animals looking for a free meal. Succulent green plants are an obvious target for their appetites. While the considerable regenerative capacity of plants permits them to overcome the loss of some leaves, a branch, or even a terminal bud, the losses inflicted by predators and herbivores can be life-threatening. Plants that have survived in the wild over the course of evolution have therefore developed a series of countermeasures to deter or repel attacks by other organisms.

Some of these countermeasures consist of obvious physical deterrents: sharp thorns on plants like the thistle discourage grazing animals by the pain they inflict, while leathery-leaved bracken ferns are difficult to bite and chew and ill-tasting when ingested. It is not unusual, in fields being grazed by cows or sheep, to see islands of thistles and bracken standing tall and untouched on a carpet of heavily grazed grass. By contrast, the ill-smelling skunk cabbage deters predators by its noxious taste and odor, most apparent after the leaves have been cut or bruised.

Even less obviously well protected plants do not suffer placidly the attacks of insects. Plants commonly interpose a complex series of structures and substances that can interfere with an insect's feeding. First, the surface of the leaf may itself be inhospitable to the insect; it may have a

Stinging hairs on a leaf of *Urtica,* a stinging nettle.

thick cuticle or a hard structure that is difficult for insect jaws to penetrate. Alternatively, the surface may be sticky, excessively hairy, or even too shiny and slippery for a firm insect foothold. Such structural features of the surface may preserve the leaf or even the entire plant from insect attack. But even if the insect penetrates this outer line of defense, it must still face an arsenal of noxious chemicals maintained within each plant cell.

Some leaves contain astringent materials like alkaloids, phenols, or tannins whose taste alone repels the insect. Other plants contain aromatic terpenes (a component of lemon oil is one example), whose penetrating odor repulses insects even before they land on the plant. Since these compounds tend to be abundant in easily broken glandular hairs on the leaf surface, any insect that does land on the leaf will receive a heavy dose, and will probably never penetrate the leaf blade itself. *Pyrethrum,* a relative of *Chrysanthemum,* produces a complex terpene in its leaves and flowers that is lethal to many insects. The dried parts of *Pyrethrum* are the basis for a natural insecticide that has become important in commercial agriculture. Since it is of natural origin and is easily biodegradable, it is preferred by many people who wish to avoid spreading noxious synthetic pesticides that have a long residence time in soil and water and thus pollute ecosystems and the food supply.

Some plant hairs are more sophisticated in their structure and function: they act like injection syringes, squirting noxious materials directly into the cells of an invader. For example, *Urtica urens,* the stinging nettle, squirts a mixture of acid, histamine, and other substances that irritate the skin and cause reddening, itching, and swelling. Any animal receiving such a mixture subcutaneously will probably cease feeding and retreat. Antihistamine compounds, which we use to shrink our mucosal membranes when we have a cold, can also be used to lessen the irritation produced by nettles. But even antihistamines cannot provide relief against the irritating oil urushiol, produced by leaves of poison ivy and poison oak, both members of the genus *Rhus.* This highly noxious material, which causes humans and other animals to give these plants a wide berth, is an irritating phenol, one of a class of slightly acidic substances that contain a phenol ring (a ring of six carbons to which a reactive OH group is attached). In urushiol, the phenol group is linked to a long chain of carbon atoms that makes the compound more lipid-soluble, and thus more able to pass through membranes. Urushiol is, paradoxically, closely related to the flavorful component of ginger, which in small doses is attractive to humans and some animals.

Sometimes, the substance noxious to insects is not present in a free, active form in plant cells, but is sequestered in complex molecules from which it must be released by the action of an enzyme. In the cell, the enzyme is contained in the cytoplasm, where it is separated from the toxin-containing material in the vacuole or other cellular compartment bounded by a membrane. The toxin is liberated only after the insect bites the leaf and chews enough tissue to rupture membranes and bring the enzymes into contact with the complex, toxin-containing molecules. Lethal cyanide gas is released in this manner from complex molecules in the nut of the almond, and noxious hydrogen sulfide (the odor of rotten eggs) and other sulfur-containing gases are similarly released from mustard oils. These compounds can sometimes kill an invader; at the very least, they discourage further browsing.

A far more subtle mechanism is used by plants that form chemical toxins only in response to injury. The leaves of such plants are toxin-free when uninjured, but form the toxin immediately after injury. An uninjured plant is thus spared the waste of energy and chemicals resulting from synthesizing a possibly unnecessary toxin. One toxin formed in this way inhibits the action of the protein-digesting enzymes of insects. After in-

Thorny protection on *Cirsium vulgare,* the spear thistle.

gesting this toxin, the insect is incapable of profiting from any of the plant's proteins that it has managed to swallow, since these will remain undigested in its gut.

The synthesis of this toxin provides an example of "action at a distance." A substance called PIIF (*proteinase inhibitor initiating factor*) is formed at the scene of chewing injury and either migrates itself or induces another substance to migrate to a distant receptor. There the mobile messenger catalyzes the formation of the toxic inhibitor by "turning on" previously inactive genes. The mobile messenger may be a peptide—that is, a small chain of linked amino acids. Such messengers have recently been found to act in animals at incredibly low concentrations; they have been hypothesized to exist in plants as well, as components of signaling mechanisms. Because of its potential importance in making plants resistant to insects, PIIF is being investigated vigorously.

Plants practice another subtle form of chemical warfare against predators: the synthesis of "unnatural" amino acids, at least unnatural for the predator although quite natural for the host plant. For example, the jackbean, *Canavalia ensiformis,* contains an unusual amino acid appropriately named canavanine, which resembles the common amino acid arginine except for the presence of an extra oxygen. If a canavanine molecule is ingested by an insect and incorporated into its protein, the protein thus formed cannot function properly and the insect dies. It is not clearly understood why canavanine is not toxic to the plant that makes it. Many other unusual amino acids are found in higher plants, and it is supposed that they function in the same way as canavanine.

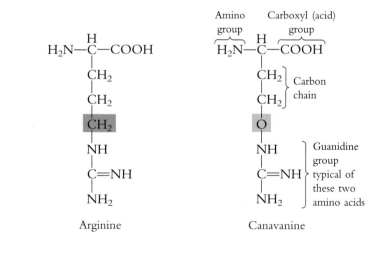

Canavanine, made only in the jack bean, differs from the common amino acid arginine only in the substitution of an oxygen for a —CH$_2$ group in the chain.

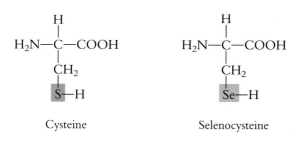

Cysteine   Selenocysteine

The toxic amino acid selenocysteine, produced by the locoweed, is identical with the common amino acid cysteine except for the substitution of a selenium (Se) atom for the sulfur atom (S).

In the leguminous genus *Astragalus,* certain plants called "locoweeds" have the unusual property of accumulating large quantities of the element selenium, which they take up from the soil. *Astragalus* plants incorporate selenium into their amino acids in place of sulfur, an element that resembles selenium in many respects. When a cow or other animal grazes upon such a plant, it breaks down the ingested proteins, releasing the selenium-containing amino acids, which are then incorporated into the animal's own proteins. Unhappily, a bovine protein constructed with the selenium-containing amino acids cannot assume the correct final shape required for the protein to function as an enzyme or structural component. As a result the cow develops a behavioral aberration referred to as the "blind staggers." It wanders around aimlessly, staggering as if drunk, and eventually develops hemorrhages, collapses, and dies. The only way for the cattle rancher to avoid this problem is to rid the grazing region of *Astragalus* plants of the type that can become locoweeds. Again, it is not known why the selenium that they take up from the soil is not toxic to the plants that accumulate it.

Although tricks based on "chemical warfare" can damage or kill insects that seek to consume a plant, they act so slowly that the plant under attack would suffer serious damage by the time a countermeasure became active. But it also appears that insects evolving over time tend to avoid the plants possessing these effective countermeasures. We must assume that some insects can sense, through smell or other means of perception, the danger to themselves from eating the plant in question. Thus, the ability of the plant to damage an insect influences the course of insect evolution. Since an insect that fails to perceive and avoid the danger will almost certainly die, it will tend to be replaced by more perceptive varieties.

Many plants destined for human consumption have been bred from ancestors that possessed chemicals or structures that protected them against predators. But over the course of many generations some of these mechanisms have been bred out. Bitter and astringent compounds deterred insects from eating the leaves of the tomato's wild ancestor. After these unpleasant compounds were deliberately bred out of the plant so that they would not interfere with the taste, the succulent and juicy tomato became a most desirable meal for insects and humans, alike. Nowadays, growers must spray back onto the tomato plant synthetic compounds that are noxious or toxic to insects. We have recently become aware that some of these synthetic compounds biodegrade very slowly; they can become incorporated into the animal food chain and in various subtle ways may cause birth aberrations, cancer, nervous diseases, or other illnesses. The desire for healthier food has spurred efforts to resort to more "natural" defensive compounds.

Recently, through the techniques of genetic engineering, scientists have sought to reintroduce the natural compounds present in the original strains of modern food plants. Thus, if one could locate the ancient ancestor of the modern tomato, then identify and clone the gene responsible for the synthesis of protective compounds, one might create a transgenic modern plant that contained the ancestral protective gene. It appears that

Both *Arabidopsis* plants have been exposed to the bacterial pathogen *Pseudomonas syringae*. The plant at the right received a pretreatment with a chemical that made it resistant to the bacterium, while the susceptible plant at the left received no such treatment. Certain genes can also confer this type of resistance, by producing chemicals toxic to the invading organism.

this can be done without producing a plant that tastes so bad that people will refuse to eat it. Ideally, similar techniques could be used to reintroduce into modern crop plants many of the protective features that have been removed over generations of breeding. These new practices are described in Chapter 8.

Fungal Invasion

Fungi are plantlike organisms that, lacking chlorophyll, must obtain their food by absorbing it from some external source. Some fungi, like the familiar mushroom, absorb their food from dead, decomposing matter such as leaves and fallen wood in the forest. These saprophytic fungi send their threadlike filaments, called hyphae, through the soil. When the hyphae encounter organic matter on which the fungus can feed, they excrete digestive enzymes that convert polymeric materials like cellulose and lignin into smaller molecules. These are absorbed and either used as an energy source or incorporated directly into the fungal hypha. At a certain point in the life cycle of the fungus, the hyphae (collectively called a mycelium) come together and fuse to form more solid organs, such as the spore-bearing mushrooms that we eat. The gills of the mushroom produce spores that can remain dormant under adverse conditions, then germinate when conditions improve to produce new hyphae, starting the life cycle over again.

Other fungi are parasitic; these fungi derive their food from living hosts. The green mold we see on rotting oranges is the spore-bearing stage of *Penicillium digitatum,* a fungus that feeds on ripe fruit. (To deter other microorganisms, especially bacteria, from taking the food that it is preparing to absorb, *Penicillium* secretes an antibiotic, called penicillin, whose discovery by Alexander Fleming in the 1940s revolutionized medicine.) Some fungi invade animals, including human beings; one familiar example is *Trichophyton interdigitale,* the fungus that is responsible for athlete's foot.

Parasitic fungi invade a plant through open stomata, root hairs, wounds, or weak surface cells. Although the fungal mycelium grows at first through intercellular space, it must eventually penetrate into living cells in order to spread from the initial site of infection. To gain access to plant cells, the mycelium pierces the nonliving cell wall surrounding each plant cell. The primary cell wall is made predominantly of cellulose and the slight variant called hemicellulose, although it may also contain lignin.

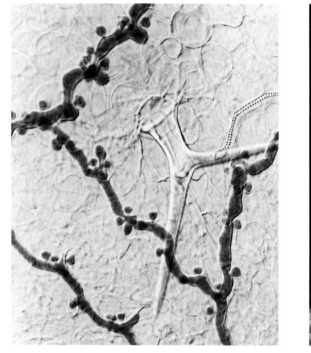 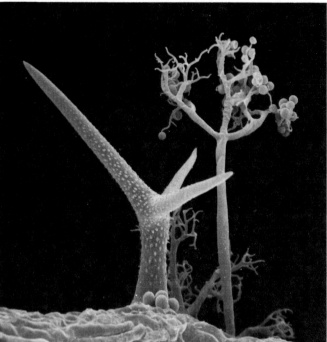

Left: The dark strands in this micrograph are fungal threads growing through cells of a leaf. *Right:* A fungal spore-bearing organ emerges from a leaf surface on the right. The structure to its left is a hair.

The walls of neighboring cells are fused together by the intercellular cement pectin. (Pectin is the major component of jelly made from fruits; it is extracted from the plant cells by boiling, and gels upon cooling.) The fungal mycelium secretes enzymes of two types that partially digest the cell walls: pectolytic enzymes act by breaking down pectin, and cellulolytic enzymes act by digesting the primary cell wall. Both enzymes cause the release of small fragments of cell wall that act as *elicitors* of considerable chemical change in plants.

Elicitors can induce the formation of special enzymes that catalyze the synthesis of compounds able to counter the growth of the fungi. One of the enzymes most commonly induced—either by injury, fungal invasion, or environmental signals such as light absorbed by phytochrome—is phenylalanine ammonia lyase, generally referred to as PAL. This enzyme has the important function of generating phenolic acids from the common amino acid phenylalanine. In the basic reaction, phenylalanine loses ammonia (NH_3), giving rise to the compound cinnamic acid, called a C_6-C_3 compound because it consists of six carbon atoms joined in a benzene ring to which are attached three carbon atoms in a side chain:

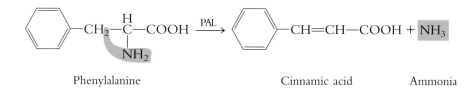

Phenylalanine   Cinnamic acid   Ammonia

A variety of other compounds can be formed from C_6-C_3 compounds by adding on groups of atoms or by linking together C_6-C_3 units.

Among the C_6-C_3 compounds that arise in plants following fungal infection are the so-called phytoalexins, which deter new infection and the growth of fungi already within the plant. The varying effectiveness of these antifungal agents is illustrated by the broad bean, *Vicia faba,* which produces a phytoalexin when attacked by the fungus *Botrytis*. The fungal species *B. cinerea* is susceptible to the phytoalexin, and the action of this chemical keeps the fungus contained in small, brown areas of dead tissue on the leaf. The appearance of the infected tissue gives the disease its name, "chocolate spot." By contrast, the related fungus *B. fabae* is able to metabolize and destroy the phytoalexin before spreading systemically through the plant.

The related bush bean, *Phaseolus vulgaris,* produces a different phytoalexin, phaseollin, in response to invasion by *Colletotrichum lindemuthianum,* the "anthracnose" fungus. Not all phaseollin-containing varieties of *Phaseolus vulgaris* are equally resistant to the fungus, and even within resistant plants not all cells are equally resistant. The susceptible infected cells of certain *hypersensitive* varieties tend to die quickly; as a result, the fungus is walled off in the dead area, while adjacent resistant cells remain uninfected. Because the walled-off fungus cannot spread to the healthy tissue in hypersensitive plants, the plant survives to produce a crop. This desirable hypersensitive behavior is now being bred into commercial crop varieties. Moreover, since most phytoalexins are able to inhibit several related fungi, the phytoalexin formed when one fungus invades a plant may protect the plant against other fungi. Some breeding programs seek to increase the overall content of phytoalexins in plants to make them more resistant to fungal invasion in general.

Competition with Other Plants

The chemical warfare so successful against predators is sometimes turned against other plants. In the wild, plants must compete for space, both for their roots in the soil below and for their leaves in the canopy of vegetation above. To stop the roots of competitive neighbors from invading their turf, some plants of the harsh desert, where close neighbors competing for water can be lethal, have evolved specific chemical warfare agents. For example, the leaves of the brittlebush (*Encelia farinosa*) contain an aromatic compound that deters the germination of seeds and the growth of seedling roots. Thus, when *Encelia* sheds a few leaves around its base, it carves out for itself a bit of *Lebensraum,* a free space in which no competitor is likely to survive.

The rubber-producing shrub of the Sonoran desert, guayule (*Parthenium argentatum*), produces a compound in its roots called parthenyl cinnamate. This compound is degraded at broken surfaces, releasing cin-

The brittlebush, *Encelia farinosa,* in bloom in the Sonoran desert. A space surrounds each plant, partly because a compound in fallen leaves has deterred the growth of competitors.

namic acid, which is toxic to guayule seedlings. In this way, the parent plant protects itself against encroachment by its own progeny. One of the most effective chemical warfare agents of all is juglone, a substance produced by the fruits of the black walnut tree (*Juglans nigra*). This compound prevents the growth of any grass or herb on the ground under a black walnut tree, thus preserving water and minerals for the exclusive use of the tree. The number of antiplant chemicals produced by plants is large, and many compounds remain to be discovered.

Responses to Injury

No defensive system works perfectly against all predators, so even well-protected plants are successfully attacked on occasion. Plants have therefore evolved not only "strategies" for defense but also strategies for surviving injury. Plants are able to repair individual wounds, and when massive damage threatens a plant's very life, the plant may "shut down" to avoid further harm. Both responses require that the plant react to signals delivered by chemical messengers.

An injured plant almost always releases a burst of ethylene from the wound area. The injury induces special enzymes to form that transform the common amino acid methionine to a substance called aminocyclopropylcarboxylic acid (ACC), from which ethylene is subsequently released. When ethylene meets a wounded cell, it activates cell division, leading to repair of the wound.

Sometimes the simple repair of a wound is followed by a further modification of the plant that protects it against injury. A good example is what happens when we peel a potato tuber, removing the surface periderm, or corky tissue, that waterproofs and protects the interior storage tissue. Peeling the potato causes the release of ethylene, which stimulates cells at the injured surface to divide and ultimately to form a new corky periderm. After some hours, the potato is covered by a thin film of new scar tissue that in a day or two will closely resemble the corky layers that originally protected the internal cells of the tuber. In a sense, the puncture of the tuber's protective seal has set in motion a chain of events, mediated by ethylene, that automatically repairs the damage.

The heady floral fragrance of some plants originates in part from a group of volatile molecules that induces an even more radical response to injury. These are jasmonates, a family of recently discovered molecules produced in some plants after wounding. Jasmonates have their origin in the breakdown of α-linolenic acid, a fatty acid found in membrane lipids; presumably, the rupture of a membrane by wounding brings together the linolenic acid and the enzyme that acts on it. Jasmonates are found widely, especially in young plant organs, from which they are commonly released in the volatile form of methyl jasmonate.

When a jasmonate is applied to a plant, it has two prominent effects. The first is to shut down many active growth processes, including pollen and seed germination, callus and root growth, and chlorophyll synthesis. The second is to promote the formation of ethylene, which favors the onset of abscission and dormancy. Because dormant, nongrowing tissues are generally more resistant to injury than succulent, rapidly growing tissues, this series of effects is probably best viewed as a complex of protective processes designed to minimize further injury to the plant. Jasmonates can also elicit a few positive responses, such as root initiation, tendril curling, and the biosynthesis of carotene, but these might all be secondary effects of ethylene release. The pattern of a plant's response to jasmonates varies widely among species.

Like many other plant hormones, jasmonates probably act by regulating the activity of genes. When researchers have sprayed jasmonates on plants, they have afterward noted numerous novel proteins, including proteins stored as reserve foods in seeds, proteins that form the structural backbone of membranes, and proteins that inhibit the breakdown of other proteins. These new proteins can change the overall pattern of the plant's metabolism, so that it is better able to resist stressful conditions. Some of these new proteins, and the genes controlling their synthesis, can be turned on by other signals as well, but the extremely low concentration of jasmonates required indicates that they are the major effective signals in the plant.

Sometimes jasmonates are produced spontaneously, without wounding. For example, an uninjured sagebrush (*Artemisia tridentata*) plant releases enough jasmonates in volatile form to induce nearby plants to change their growth habits in various ways. For example, plants incubated with the sagebrush in a sealed chamber synthesize a substance controlling protein breakdown. This ability of sagebrush to control the growth processes of its close neighbors may have ecological significance, in that it could give the plant a competitive advantage in the struggle for space in its arid habitat.

A proposed scenario for jasmonate intervention in the wounding response is as follows: injury breaks apart cellular membranes, which then come into contact with enzymes that partially digest the membrane frag-

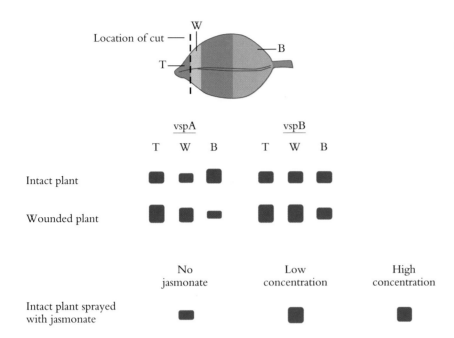

Location of cut

vspA   vspB

T   W   B   T   W   B

Intact plant

Wounded plant

No   Low   High
jasmonate   concentration   concentration

Intact plant sprayed
with jasmonate

Top: When a leaf was wounded by cutting off its tip, it produced more mRNA for the genes *vspA* and *vspB* in the cut-off tip (T) and in the area near the wound (W), but not at the base of the leaf (B). *Bottom row:* Intact plants sprayed with only a low concentration of jasmonates also produced much mRNA for the *vspA* gene, suggesting that jasmonates may activate this gene after wounding.

ments. The fragments release free fatty acids, including α-linolenic acid. Other enzymes act to transform α-linolenic acid to other compounds, including jasmonates. Jasmonates then control the action of genes, turning some on and others off. The resulting new spectrum of proteins, including enzymes, produces the wounding response. It is interesting that jasmonates and related compounds also function as sex pheromones for insects and control reproductive activity in some fungi. Jasmonate research will continue to flourish in the years ahead.

Excessive Salt

The pressure to produce more food for the world's ever-growing population has led to the irrigation of many acres of arid soils. All the water used for irrigation has some salts dissolved in it; as water evaporates from the soil, the salt is left behind. Because plants absorb salt much more slowly than they do water, the salt content of arid soil under irrigation tends to rise constantly. As the salt content of irrigation water itself also rises over time, the salinity of water in the soil becomes so great that plant growth suffers.

Heavy incrustations of salt coat the surface of Tamarisk *(Tamarix pentandra)* leaves collected in the Mojave Desert, California. The incrustations were created when a concentrated brine excreted from salt glands crystallized.

Salinity damages plants for two reasons. First, the salt concentration in soil water becomes so high that the uptake of water becomes increasingly difficult for purely osmotic reasons. In extreme cases, the salt content of the soil water may exceed that of the plant cell, so water may actually tend to leave the cell osmotically, rather than enter it. This effect is the basis for the use of salt to kill unwanted trees or shrubs. A second, and probably more important, reason for salt damage is that ions such as sodium are toxic. Plant membranes function well when they are in contact with a balanced salt solution. When they are surrounded by a solution of a single salt such as sodium chloride, they become leaky, and the cell must work hard to exclude sodium from the cell, using the energy of respiration. If a little potassium chloride is added to the sodium chloride solution outside the cell, the toxic effects of the sodium ion decline markedly. There is also a requirement for balance between monovalent positive ions such as sodium (Na^+) and potassium (K^+), which are missing one electron from the atom's full complement, and divalent positive ions such as calcium (Ca^{2+}) and magnesium (Mg^{2+}), which are missing two electrons. Thus, a cell can tolerate a balanced salt solution much better than it can a single salt.

Some plants (glycophytes) are very sensitive to salt, others are salt tolerant, while still others (halophytes) may even grow best in saline environments. It is not unusual for halophytes to be stimulated in their growth by sodium concentrations that would kill many glycophytes. Plants make use of several tricky adaptations when they assume the halophytic way of life. Some halophytes exclude ions such as sodium at the membrane, using the energy stored in ATP. Others absorb sodium, but sequester it in vacuoles or other special membrane-bounded cell compartments, where it is not free to influence the chemistry of the cell. Then, to compensate for the osmotic effects of these salt-laden compartments, the plant synthesizes organic substances such as the amino acid proline, whose concentration in the cytoplasm rises to levels high enough to counter the osmotic "pull" for water exerted by the saline vacuoles. The presence of proline thus permits the cytoplasm to remain appropriately hydrated.

Some unusual plants absorb salts from saline groundwater through their roots, transport the salt through the xylem to leaves, and then get rid of much of the salt by excreting a concentrated brine through special salt glands. The plant must expend energy to concentrate the salt for excretion, since it has to "pump" salt out through the membrane to counter the tendency of randomly diffusing ions to equalize their concentration everywhere. The pumping is carried out by special membrane proteins whose synthesis is coded for by genes activated by high salt concentrations. Some halophytic plants, like the desert plant *Atriplex* (the salt bush) and the Mediterranean tamarisk shrub (*Tamarix*), are virtually covered with salt crystals formed from the extruded saline solutions.

The plant seems to have learned well the lessons of adaptation in its fight for existence under unfavorable conditions. Survival mechanisms remain encoded in the genes of successful wild plants, and people have learned to use some of these tricks to breed or to protect successful cultivated plants. In this instance, human wisdom derives from "green wisdom" built up over eons of evolution.

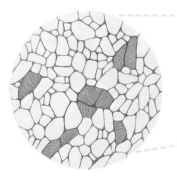

We are told that important scientific discoveries are sometimes made in clusters. One familiar example is the rediscovery in the year 1900 of Gregor Mendel's findings in genetics, approximately 35 years after their original publication, by three prominent scientists: E. Von Tschermak in Prague, Carl Correns in Germany, and Hugo De Vries in the Netherlands. This phenomenon is frequently "explained" by the statement that the time was ripe for a particular discovery or that some generalization was in the air and available for those who would grasp it.

Whatever the cause, in the early 1930s experiments were begun in several different laboratories that led to the discovery of a remarkable ability of plants: under the right conditions, their cells can continue dividing almost without limits. From this ability comes the power of plants to regenerate lost organs; indeed, the entire plant can frequently be regenerated from even a single cell of a root, stem, or leaf. In this regard, plant cells contrast with most animal cells, which have a limited ability to divide. For example, Leonard Hayflick found that human cells grown in artificial culture can divide about 50 times until they stop, for some yet unexplained reason; only transformation to a cancerous state prevents this otherwise inevitable cessation of growth.

One of the pioneers of the 1930s was Philip R. White, an American botanist who attempted to grow root systems cut off from the rest of the

SIX

Regeneration:
From Cell to Plant

plant. These roots were separated from their normal supply of nutrients, which in an intact plant are delivered to the roots by the vascular system of the stem; therefore, White had to supply the nutrients himself. He excised various root apices and placed them in a salt solution containing basic nutrients and a source of carbon such as sucrose. As the excised root grew, he periodically cut off 1 cm or so of the root at its apex, containing the meristem, and placed the newly cut-off piece in fresh medium in a new flask. The root apices grew vigorously for a while, but they gradually stopped growing during successive transfers. White discovered that by adding certain rich nutrient sources such as yeast extract to his basic solu-

1

Seeds are
sterilized in
sodium hypochlorite.

2

They are set on agar
to germinate.

3

The seeds germinate
in about two days.

4

Seedlings are separated,
and the tip of the root
is excised.

5

The root tip is placed
in a nutrient medium.

6

It grows into a large root,
from which the tip may
be again removed.

The culture of potentially immortal root tips. The cells of the organized root apex give rise to regular root tissue in culture.

tion, he could greatly improve the growth of his excised roots; they generally grew faster and through more transfers. Analyzing the separate known components of yeast extract (which were just becoming understood), he was able to demonstrate that the addition to the culture of thiamine, or vitamin B_1, greatly extended the active life of his excised root system. In some instances, the addition of thiamine conferred potential immortality on the excised roots; that is, successive subcultures of the freshly excised tip, when transferred to fresh medium, continued to grow at an undiminished rate as long as infection was prevented.

In their experiments, White and others like him were not only discovering some unusual properties of plant cells; they were also perfecting one of the century's most important experimental tools—tissue culture. By identifying the substances, known as growth factors, required for the continued flourishing of plant cells, they made possible the growth of groups of cells outside the plant, where the cells could be observed and manipulated in experiments to an extent previously undreamed of.

While White was experimenting with various root tips, William J. Robbins at the New York Botanic Garden was investigating the growth of excised tomato roots. He was able to confirm White's findings that thiamine was essential for indefinite growth, but in addition he discovered that his tomato roots required a second nutrient to continue growing. This nutrient was pyridoxine, also called vitamin B_6. At about the same time, James Bonner, working at the California Institute of Technology, was cultivating excised pea roots. Again, thiamine turned out to be essential for indefinite growth, but in addition the roots required nicotinic acid (niacin), yet another member of the vitamin B group. Now, many decades after this pioneer work, we know that virtually all excised root systems require thiamine for their continued growth, and sometimes another substance as well. The second substance may be any of a variety of nutrients, and sometimes turns out to be unnecessary in the long run. For example, Philip White routinely added the amino acid glycine to his cultures, but most modern cultures omit this substance and the roots continue to grow successfully.

In Europe, Roger Gautheret, working in the laboratories of the Sorbonne University in Paris, was investigating the growth of cubes of fleshy carrot root. Gautheret sterilized the surfaces of his excised cubes and transferred them to sterilized media, which contained basic nutrients supplemented with sucrose and sometimes with other organic nutrient sources, such as yeast extract. After incubation for a month or two, some of these cultures showed a surface layer of fuzzy white cells. Experimenting with the then newly discovered growth hormone auxin, Gautheret

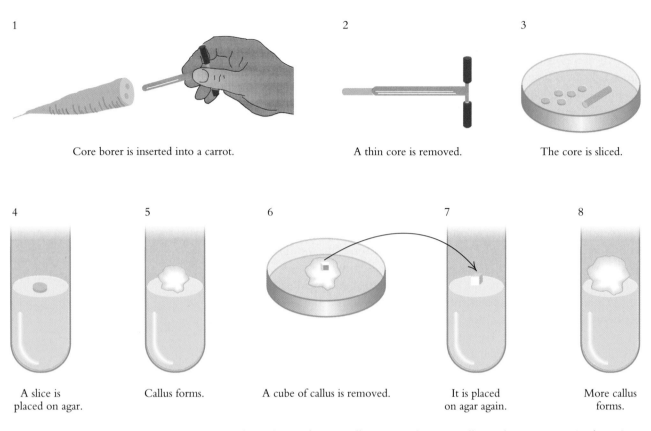

1

Core borer is inserted into a carrot.

2

A thin core is removed.

3

The core is sliced.

4

A slice is
placed on agar.

5

Callus forms.

6

A cube of callus is removed.

7

It is placed
on agar again.

8

More callus
forms.

The culture of potentially immortal carrot cells. Lacking an organized meristem, the explant produces undifferentiated callus tissue.

found, to his satisfaction, that by adding auxin to the medium he could
greatly increase the rate at which his cultures produced the new white
cells.

Gautheret's careful experiments revealed that cells from different parts
of the root required different amounts of auxin; thus most carrot cultures
required as little as 10 parts per million of auxin, while any culture containing a root apex required no auxin at all, and other cultures seemed to
require intermediate quantities. Gautheret reasoned that all carrot root
cells growing in culture required some auxin, but that apices produced
enough auxin to satisfy the needs of the entire culture. In addition, certain
cells from outside the apex possessed or developed the ability to synthesize
some auxin, although in insufficient amounts. He also discovered that
some cells that originally required the addition of auxin became "habitu-
ated" after some time in culture, thereafter requiring less auxin or no
auxin at all. This mysterious phenomenon is still not well understood.

There is an important distinction between the results of Gautheret, who cultured pieces of fleshy roots without their apices, and those of previous investigators, who cultured roots with the apices. Excised root apices, bearing regularly aligned cells organized into functional tissue masses, gave rise to cells that became organized into a similar pattern; in other words, organized meristematic root produced organized root. However, the excised tissue in the Gautheret experiments contained no organized meristem to guide development; as a result, the culture produced undifferentiated new tissue, called callus tissue. The cells produced on the surface of the excised cubes of carrot tissue were not at all like the cells from which they were derived. Rather, they were unpigmented, thin-walled, and irregular in shape; they were not aligned in neat rows as were their predecessors nor were they arranged in any other well-organized pattern. Clearly, although excised root cells are able to reproduce when supplied with the appropriate growth factors such as thiamine or auxin, the form of the cells produced is determined by some other characteristic of the tissue. Whatever the characteristic is, it is more likely to be found in well-organized meristematic tissue.

The undifferentiated mass of white fluffy cells in the callus tissue maintains its rate of growth indefinitely, as long as the tissue is transferred frequently to fresh, sterile medium. Callus cells are thus potentially immortal: they continue to survive and reproduce as long as they are sup-

Tracheids

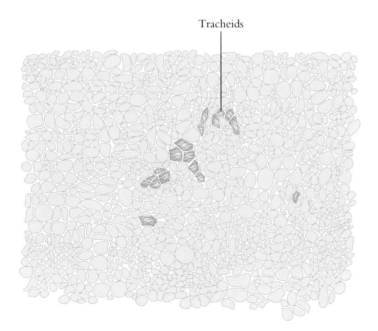

A cross section of a carrot callus shows mostly undifferentiated cells and a few nonfunctional tracheids.

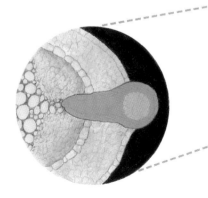

Microorganisms are found everywhere in the biosphere; indeed, these opportunistic organisms have colonized virtually every available ecological niche in nature. On the ocean floor, they make a bizarre living by oxidizing iron and sulfur released from volcanic vents; their ability to metabolize these elements has made them the ultimate source of nutrients and energy for an elaborate animal community. Deep within the soil, they use the energy gained by the oxidation of organic matter to secrete acids that degrade even feldspar and quartz, thus aiding in the formation of soil from rock. They range as high as the highest mountain tops, and their spores can be found in the upper reaches of the atmosphere. Because of their ubiquity, it is not surprising that they should frequently encounter higher plants, and because of their metabolic versatility, it is not surprising that they should interact extensively with them.

Much of the surface of the plant body is covered with a film of bacteria and other microorganisms. For example, the growing zone at the tip of the root synthesizes organic molecules of various kinds needed for root growth; some of these molecules leak or are excreted into the surrounding soil (the rhizosphere), where they attract motile bacteria and other microbes, including filamentous fungi. These organisms feed on the ex-

There is an important distinction between the results of Gautheret, who cultured pieces of fleshy roots without their apices, and those of previous investigators, who cultured roots with the apices. Excised root apices, bearing regularly aligned cells organized into functional tissue masses, gave rise to cells that became organized into a similar pattern; in other words, organized meristematic root produced organized root. However, the excised tissue in the Gautheret experiments contained no organized meristem to guide development; as a result, the culture produced undifferentiated new tissue, called callus tissue. The cells produced on the surface of the excised cubes of carrot tissue were not at all like the cells from which they were derived. Rather, they were unpigmented, thin-walled, and irregular in shape; they were not aligned in neat rows as were their predecessors nor were they arranged in any other well-organized pattern. Clearly, although excised root cells are able to reproduce when supplied with the appropriate growth factors such as thiamine or auxin, the form of the cells produced is determined by some other characteristic of the tissue. Whatever the characteristic is, it is more likely to be found in well-organized meristematic tissue.

The undifferentiated mass of white fluffy cells in the callus tissue maintains its rate of growth indefinitely, as long as the tissue is transferred frequently to fresh, sterile medium. Callus cells are thus potentially immortal: they continue to survive and reproduce as long as they are sup-

Tracheids

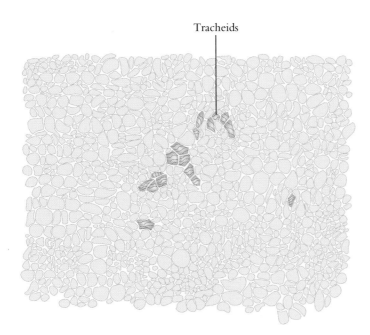

A cross section of a carrot callus shows mostly undifferentiated cells and a few nonfunctional tracheids.

plied with nutrients and growth factors. If one makes a section through the cauliflower-like mass of a callus, one sees mainly undifferentiated thin-walled parenchyma cells, but occasionally, deep within the tissue, one sees highly lignified tracheids, the thick-walled cells that in the vascular system conduct water and minerals. These tracheids are oriented as if they were part of a conducting system, but they do not go from any logical source of nutrients to any set of cells requiring nutrients. Such vascular strands containing tracheids are generally short, nonfunctional, and not nearly as well aligned as in normal tissue.

Gautheret and his student, Guy Camus, later showed that they could cause the undifferentiated cells in a callus to take on a well-structured form. All they had to do was connect the callus to organized differentiated tissue, such as a bud. For example, Gautheret and Camus transferred a terminal bud of lilac into an undifferentiated mass of callus tissue. After some time, cells at the cut base of the bud grew and fused with cells of the callus; then, some influence from the bud, partly replaceable by auxin, was transmitted into the callus tissue, so that the parenchyma cells underneath the vascular strand of the bud were transformed into functional vascular tissue, mainly tracheids. Thus, tissue organization begets organization, while disorganization left to itself produces mainly disorganization, although occasional random events produce differentiated, organized cells.

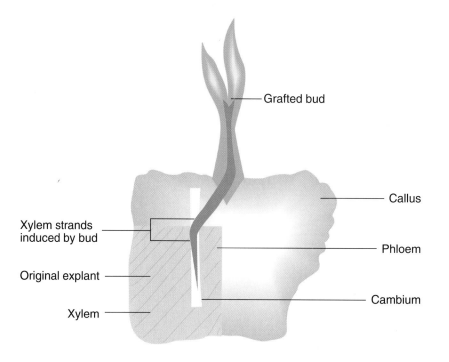

When an organized bud is grafted onto a partly undifferentiated callus, new vascular cells form to connect the donor and receptor.

The Totipotency of Single Plant Cells

How small a piece of, say, carrot root can one transfer to a culture and still expect the tissue to regenerate? In experiments directed along these lines, Gautheret found that a greater percentage of transfers were successful when the cubes were rather large. As he progressively decreased the size of a face of the excised cube, the chances of regeneration declined, and below a few square millimeters cell reproduction came to a halt. Gautheret reasoned that the cell sap liberated from the wounded area was injuring the carrot cells and hindering their reproduction. Such sap would include noxious materials such as phenols, normally stored in the central vacuole, as well as toxic substances formed when previously sequestered enzymes were brought into contact with cellular components.

The cells in small pieces of carrot root might still regenerate, however, if none were damaged. Although it is difficult to isolate small assemblages of cells without creating wounds, individual cells can be obtained without wounding or cutting by placing a callus culture onto an oscillating shaker. Most of the cells remain "glued" together by pectic materials lying between adjacent cells, but some cells break free uninjured and stay suspended in the culture fluid. An individual cell may be removed and placed in a large volume of new medium, but under these conditions it generally does not proliferate. However, when many cells of the same kind are seeded into a small area, some of them develop.

After many investigators had repeated experiments of this kind, the idea arose that the medium had to be "conditioned" to support the growth of individual cells. It appeared that cells placed into fresh medium leak out certain materials such as vitamins, hormones, or amino acids essential to their growth and replication. When a single cell was placed in a large volume of medium, its loss of such essential factors was so great that growth was impossible. When, however, many such cells were placed in the same volume of medium, their collective leakage created a pool of nutrients adequate for growth. From such experiments also came the idea that cells could be seeded into previously conditioned media—media that had been in contact with "leaky" growing tissues but whose nutrients had not been exhausted by such contact. In a variant of this technique called the nurse tissue or feeder layer technique, a single cell is placed in contact with, but is not allowed to fuse with, a lower layer of cells in contact with the medium. The growth medium diffuses through the layer of cells to the single cell, carrying with it essential materials drawn from the growing cell layer.

Using a feeder layer technique with cells of tobacco pith, William Muir and Albert Hildebrandt at the University of Wisconsin were able to

Even a single cell of tobacco can divide to produce a callus, and ultimately an entire plant. Nurse tissue helps to start the growth of the isolated cell.

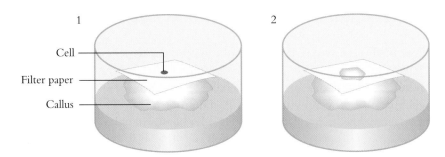

show that a single cell of tobacco can be stimulated to produce a large cellular mass. Later it was found that such a cellular mass could eventually give rise to an entire plant. These experiments provided proof that the pith cells of the tobacco plant are totipotent, which simply means that a dividing pith cell is eventually able to produce all the different cells of the mature body. The phenomenon confirms that each cell carries all the genetic instructions for creating the entire organism.

In the 1950s, Frederick C. Steward and his colleagues at Cornell University developed a successful variant of the nurse tissue technique. Their method made use of a rotating flask complex in which small bits of

Roots (visible) and embryoids (not visible) form when carrot cells are tumbled gently in glass flasks bearing nipplelike projections.

tissue were tumbled gently back and forth from air to the liquid culture medium. They found that in such an apparatus, entire plants could readily be regenerated from single cells taken from the callus cultures of carrot and of its wild relative Queen Anne's lace. The mass of cells organized itself into a spindle-shaped, bipolar structure closely resembling the carrot's normal embryo. These structures were named *embryoids,* and the pathway from single cell to embryoid to plant was called the embryogenic pathway, as opposed to the organogenic pathway, in which a callus first gives rise to a root, then a bud, then ultimately forms an entire plant. Both techniques have successfully furnished multiple plants from single cells, thus providing the basis of a new industry for the propagation of plants, called micropropagation. Since the cells to be propagated may have been altered by genetic engineering, micropropagation has become a tool whereby a deliberately engineered cell may be transformed into numerous adult plants for use in agriculture.

Protoplasts

Theoretically at least, any living cell taken from the body of a higher plant can, under appropriate conditions, divide repeatedly to form an entire new organism. Thus, the basic unit of plant regeneration can be as small as a single cell. It would seem that this is the ultimate reduction that can be achieved; yet, there is a little further that we can go in our search for the smallest unit from which an entire plant can be regenerated.

When fungi invade higher plants, they commonly cause hard, solid tissue to soften and disintegrate. By digesting the pectic cement that holds cells together, fungi can reduce a solid mass of cells into a collection of individual cells. They then secrete enzymes that digest the remaining cellulose cell wall, liberating the cell contents. When in place, the cell wall acts as a pressure jacket restraining the expansion of the cytoplasm-bounded vacuole, which otherwise would take in water through osmosis because of its high salt concentration. Once the cell wall has been digested, the vacuole continues expanding until the cell bursts. If, however, the cellulose wall is disrupted in a medium that has a higher osmotic concentration than the vacuole, the cell does not burst. Rather, it remains in the form of a spherical mass, the *protoplast,* containing all of the organelles of the cell except the cell wall.

In the laboratory, the experimenter can obtain protoplasts by suspending tissue in high concentrations of an appropriate solute, then adding purified cellulose-digesting enzymes obtained from fungi. The result is the conversion of a leaf, stem, or root into a mass of beautifully spherical protoplasts. When carefully rinsed free of the digestive enzyme and transferred to nutrient-containing medium, some protoplasts are able to form new cell walls, resume division, and regenerate an entire organism, by either the embryogenic or the organogenic pathway. Thus, the protoplast, not the intact cell, is the smallest unit of the plant body capable of regenerating an entire plant.

Naked protoplasts have made possible an entirely new technology for developing and propagating new kinds of plants, referred to as somatic hybridization. Let us assume that we isolate cells from a tomato leaf and a potato leaf, both plants of the family *Solanaceae,* and convert both groups of cells to protoplasts. Finally, we mix the two populations together. They can easily be distinguished by the shape of their chloroplasts and the type

1

A leaf is peeled,
to expose the inner cells.

2

Strips of peeled tissue are
incubated with cell wall-digesting
enzymes, and protoplasts released.

3

Protoplasts are purified
by centrifugation.

4

They are placed on
a complete medium.

7

5

Some develop cell walls, divide,
and form large callus masses.

6

Some calli develop
roots and shoots.

They form entire plants
when properly transplanted.

From plant to protoplast and back to plant. A single protoplast can produce an entire plant.

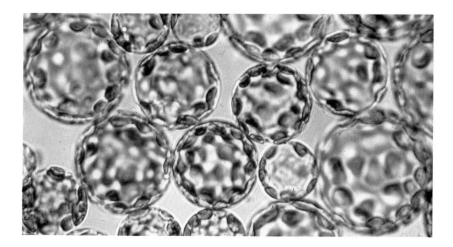

Mesophyll protoplasts have been isolated by digesting the cell walls of leaves in the presence of a high concentration of mannitol to prevent bursting.

of pigmentation. In the appropriate medium, the membranes of such protoplasts can be "melted" and caused to flow together, yielding clusters of protoplasts fused to produce a giant protoplast. Some clusters will represent only tomato protoplasts, while others will represent only potato protoplasts. But still others will represent the fusion of both tomato and potato protoplasts. By diluting the protoplasts, we can favor the fusion of only a single potato protoplast with only a single tomato protoplast. If the two fused protoplasts are closely enough related, as in this instance, they can be regenerated to form an entire plant displaying characteristics of both the parent plants (a "pomato" or "topato"). From such fused protoplasts have developed several new agricultural strains.

What might we expect from the fusion of a tomato and a potato? Would the new plant produce both edible tomato fruits and underground potato tubers? Would only one of these characteristics appear, or would neither? The answer seems to be, following the "law of maximum human unhappiness," that the new plant yields neither useful food product. We conclude that the fused genome is somehow unbalanced and unable to develop normally. However, obtaining the desired mix of characteristics is a matter of trial and error, and it is still possible that useful plants can be derived from the crude mixing of genomes.

A cluster of protoplasts can be conveniently fused by an electric "zapper"; this device lines cells up in an alternating-current field and then delivers a jolt of direct current that instantly melts their membranes. Alternatively, protoplasts may be fused by suspending them in polymeric solutes such as polyethylene glycols. These materials seem to dissolve in the membrane, linking the protoplasts by causing adjacent membranes to flow

These tubers from "topato" plants were produced by fusing protoplasts derived from a potato and a tomato.

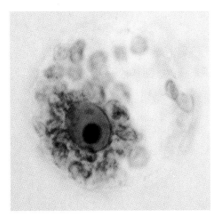

Liposomes may be used to introduce a single protein into a protoplast. Here two linked proteins have been transported into the nucleus of a protoplast. One is a "reporter" protein that produces the blue color to indicate the transfer has been successful.

together. In both methods, the membranes fuse and regeneration begins shortly thereafter, and both methods have been successfully employed to produce new types of plant cells.

The fusion of two protoplasts unites the complete genomes of both: the portion carried in the nuclei and much smaller portion carried in the chloroplasts. It is not necessary, however, for both fusion partners to carry the entire genome. For example, isolated nuclei can be obtained in bulk from protoplasts by disrupting the cells and spinning the resulting liquid material in a centrifuge. Molecules of differing mass and shape move at characteristic speeds in a centrifugal field; thus, particles of the same kind, such as the DNA-containing nuclei, will accumulate at the same position in the test tube. Once a sample of nuclei has been obtained, the nuclei can be encased in natural lipid membranes, forming spherical bodies called liposomes. A liposome can be fused with a second, entire protoplast by the same techniques that are employed in protoplast fusion. Chloroplasts may also be gathered in this way and then introduced into entire protoplasts. Thus, scientists may study the effect of various combinations of nuclear and chloroplast genomes on the expression of particular characteristics in plant cells.

Haploid Protoplast Techniques

The nuclei of ordinary cells of the plant body contain two sets of chromosomes, one derived from the sperm, the male sex cell produced in the pollen tube, and one derived from the egg, the female sex cell found in the ovule. Such a cell is said to be diploid, and to have the $2n$ number of chromosomes, or twice the number n found in a sperm or egg. When two $2n$ protoplasts are fused, they produced a $4n$, or tetraploid, cell. Some tetraploids are hardy and vigorous, but others are not. To avoid the deleterious effects that may be produced by tetraploidy, one can fuse protoplasts obtained from the $1n$ or haploid tissue, such as that producing sperms and eggs.

Several decades ago, various investigators began attempting to obtain haploid plants through the direct regeneration of haploid egg and sperm cells. In the 1940s, Johannes van Overbeek, working at Cold Spring Harbor, New York, tried to regenerate haploid tissue from the egg of the Jimsonweed, *Datura stramonium*. He failed, but while trying to make his isolated cells grow, he discovered that coconut milk and other liquid endosperms were remarkably potent stimulators of growth. About a decade later, Sipra Guha and Satish Maheshwari in India achieved success with the same plant, using immature pollen rather than egg cells. They

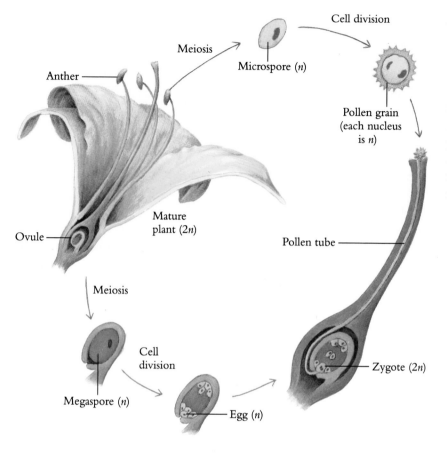

Anther

Meiosis

Microspore (*n*)

Cell division

Pollen grain
(each nucleus
is *n*)

Ovule

Mature
plant (2*n*)

Pollen tube

Meiosis

Cell
division

Megaspore (*n*)

Egg (*n*)

Zygote (2*n*)

A 2*n* cell within the ovule undergoes reduction division (meiosis) and produces four 1*n* cells, or megaspores. Only one of these survives to divide again, creating eight 1*n* cells, including the egg. Similarly in the anther, 2*n* cells undergo reduction division to produce 1*n* microspores, which divide again, eventually yielding sperm. The fusion of egg and sperm during fertilization restores the 2*n* condition.

cultivated excised anthers of Jimsonweed and found, happily, that immature pollen cells sometimes continued division and gave rise to haploid callus and then an entire haploid plant, usually through the embryogenic pathway.

Haploid plants obtained in this manner are extremely useful in breeding operations. If the roots of a haploid plant are exposed to a drug called colchicine, the absorbed drug interferes with the process of cell division: the chromosomes divide but do not sort out into separate cells. When such cells are reconstituted after division, instead of being haploid, they are diploid, carrying a duplicate set of chromosomes identical to the first set. The genes of each gene pair in such plants are identical, or *homozygous*. Such "doubled-up haploids" of cereals such as rice and wheat have been extensively used in plant breeding; because their two sets of chromosomes are identical, they carry no hidden traits to appear unwanted in

Haploid embryoids have arisen from
a cultured stamen of *Datura*.

future generations. Thus the eggs and pollen obtained from such plants must carry the original traits over to future generations.

The protoplasts of either haploid or diploid plants can provide useful systems for studying the effects of any particular gene in plant development. A gene is isolated from an external source and cloned; the cloned genes are easily introduced into protoplasts by liposomes or direct injection, or as part of a virus or a plasmid derived from the common microorganism *Agrobacterium tumefaciens*. We shall discuss the techniques involved in these manipulations in Chapter 8. If the altered protoplasts successfully regenerate a cell wall and grow in culture medium, the resulting tissues can undergo embryogenesis, producing numerous identical plantlets, which can then supply the stock for a new plant genome.

Hormones and Plant Cell Regeneration

Much scientific understanding of plant cells in culture has come from observing their reactions to applied plant hormones and other growth factors. We have noted White's observation that excised root tips need thiamine and Gautheret's discovery that the addition of auxin greatly promotes both cell division and cell extension in excised carrot callus and other tissues. We have also mentioned the experiences of van Overbeek in attempting to regenerate haploid eggs of *Datura stramonium*. Although the ovarian tissues could grow when supplied with auxin, their growth was much faster if coconut milk or some other liquid endosperm was added to the auxin-containing medium. It was this observation that led

Folke Skoog many years later to attempt to isolate the substance promoting cell division in his test material, the stem pith of tobacco. His efforts eventually led to the isolation of the hormones known as cytokinins; among their other characteristics, cytokinins are defined as substances that promote the division of tobacco pith cells (and other test materials) in the presence of auxin. So far as we know, all dividing cells require auxin and cytokinin for their division.

As it divides, an undifferentiated plant cell produced in culture can follow various pathways to produce cells of different types, depending in part on its exposure to hormones. When certain cells divide after receiving genes that cause additional auxin production, they produce calli that regenerate large numbers of roots. Such cultures, referred to as hairy root cultures, are widely employed to study the metabolism of root tissues and to produce large quantities of substances synthesized exclusively in roots. If, on the other hand, genes leading to the production of extra cytokinins are introduced into similar receptor cells, the production of buds is enhanced, and the production of roots is restrained. When genes coding for both extra auxin and extra cytokinin are introduced into the same cell, the resulting calli produce rapidly growing tumors that share certain characteristics of the malignant cancers found in animals. Thus, the rate at which growth hormones are produced by a plant cell has much to say about what path of development the cell will follow, and about what its ultimate fate may be.

Whether in nature or in cultures, cells will not begin dividing unless the hormones auxin and cytokinin are present. But once a rapidly dividing cell culture has been established, the hormones often must be removed for the cells to properly differentiate to form, for example, an embryoid. Thus, to achieve a particular pattern of differentiation, the culture may need to be exposed to hormone for a set time, then removed to a hormone-free medium, or the culture may have to undergo some other sequence of the two operations. These findings suggest that in nature, the hormone levels in tissue may vary in set sequences that are crucial to the orderly development of the plant.

Plant cells sometimes gain versatility as they form multicellular groups from isolated cells. About two decades ago, Kiem Tran Thanh Van, working at the Gif-sur-Yvette Laboratories of the French National Center for Scientific Research, was studying cultures grown from thin layers of tissue peeled from the surface of tobacco shoots. The thin strips consisted of epidermal layers plus several subepidermal layers, which are layers of compactly arranged cells directly underneath the epidermis. In a few instances, the strips included cortical tissue as well. The explant developed callus tissue, as expected, but her interesting finding was that the callus tissue could give rise to different organs, depending on whether the donor plant was in flower or not. When the strips had been removed from the surface of tobacco plants not in flower, they produced cells that could either continue growth as undifferentiated calli or that could produce roots or leaf buds. The exact developmental fate of the culture seemed to depend mostly on the auxin to cytokinin ratio, as in other systems we have examined. If, however, the surface layer was derived from a flowering plant, especially from the flowering stalk itself, then floral organs or entire flowers could be produced directly from the tissue culture, in addition to the usual leaf buds, roots, and callus. It appears that the inception of flowering in the donor plant changes the potential of its stem cells to regenerate organs in tissue culture. This is not surprising, since, as was noted in Chapter 2, photoperiodic induction changes many aspects of growth all over the plant.

Epidermal strips have also been removed from the flowering stalk of tomato plants in flower. Like strips taken from the related tobacco plant, strips from tomato plants give rise to flowers and eventually mature fruits directly from the callus without the intervention of any other vegetative tissue. The floral buds form more readily when substances that are slowly converted to auxin are added to the medium, rather than auxin itself; the slow transformation of these substances provides a steady, low level of auxin rather than one large dose.

A thin surface strip of a tomato flower stalk (*left*) produces a callus (*center*) and leaf buds (*right*).

Organ Regeneration in Injured Plants

Regeneration patterns in intact plants reflect the variety and versatility of those seen in cultured cells. Many cells that do not divide in the intact plant nevertheless retain a capacity for division that becomes apparent only when the plant is injured. The pith of the tobacco stem is an excellent example: the excised cells spring into vigorous division when supplied with sufficient auxin and cytokinin. Probably a shortage of one or both of these growth hormones limits the division of pith cells in the intact stem. Sometimes, cutting alone is enough to provide the stimulus for further growth, as when the accumulation of auxin at the base of a cut stem initiates adventitious roots, or injury to an apical bud leads to the formation of one or more lateral buds to take its place.

Another example of regeneration is supplied by the potato tuber, which is a modified stem. There are numerous buds in the axils of the scale leaves (the "eyebrows" over the "eyes" of the potato); these buds are kept dormant through the combined action of auxin from the apical bud and an inhibitor in the corky skin of the tuber. The farmer propagates potatoes not from seed, which would be too slow, but from the tuber, which is cut into about six pieces, each bearing a single eye. The previously dormant buds in the eyes sprout and give rise to an entire potato plant. This is an example not only of regeneration but also of propagation, since each tuber yields several plants. The stolons (horizontal stems) of such plants as strawberry and quackgrass can also be used for propagation, since single isolated nodes can take root to form entire plants. Thus, chopping up the quackgrass weed serves only to propagate it.

Planting stem cuttings is a tried and true method of plant propagation. Short pieces of stem bearing a few leaves are excised from the plant, and their bases are placed in water or moist rooting medium. Such leaf-bearing pieces of stem may produce enough auxin to initiate roots spontaneously at the base. Should the cutting's supply of auxin be insufficient, the base may be dipped momentarily into a solution of synthetic auxin to stimulate the growth of roots. When transplanted, these rooted cuttings will give rise to entire plants.

Regeneration may also occur from the base of certain leaves. For example, adventitious roots may arise from the petiole of the African violet (*Saintpaulia*). A single leaf excised from this favorite household plant can serve as the origin for an entire colony of genetically identical daughter plants. The leaves of the unusual plant *Kalanchöe* produce well-formed daughter plantlets spontaneously in the notches of their margins. Such plantlets produce several pairs of leaves and small roots while still on the

Many fully formed seedlings are produced at the leaf margins of *Kalanchoë*.

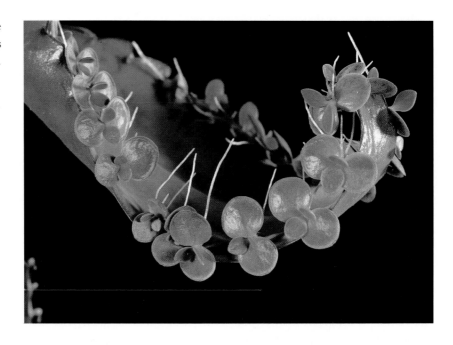

mother plant. When removed from the original leaf and placed in separate pots, these plantlets quickly grow into identical copies of the mother plant. This is an extraordinarily rapid means of vegetative reproduction.

Grafts and Chimeras

In horticulture it is sometimes desirable to graft a stem (scion) onto an alien root stock to ensure vigorous productivity or even survival. This practice is common in the wine industry, where the stems of desirable varieties are often susceptible to disease and insects; to produce a grape crop, these stems must be grafted onto root stocks that are resistant to soil-born pathogens and insects. The graft union is generally cut V-shaped or cleft-shaped to expose the maximum surface of cambial cells, whose meristematic activity must form a living bridge between the cells of stock and scion. If such cells do grow together and if vascular tissues make the proper connections, the grafted plants perform as normal single individuals, although they are composed of two different genetic types.

Sometimes, during the healing of the grafting wound, the cells of one graft partner will interdigitate with cell layers of the other partner; if buds should happen to arise from such mixtures of cells at complex graft unions, then a *chimera* may result. Chimeras are single organisms containing cells of two or more different genetic origins. Chimeras may contain pie-shaped sectors, surface layers, or even entire meristems that are divided in their genetic makeup. Because the cells of the two different partners in a chimera are often easy to distinguish by size, color, or structure, these organisms can be used to great advantage in the analysis of how plant structures develop. For example, if only the epidermal layer comes from one partner, then the role of the epidermis in the formation of the complex structure of leaves or flowers can easily be followed. If, on the other hand, an entire sector from the meristem down belongs to one chimeral partner, then the fate of the descendants of each cell in the mixed meristem can be traced. Progeny of such "marked" meristematic cells have been followed in the poinsettia for 50 successive nodes on the stem, and in the cranberry for 100 successive nodes.

A chimera, containing normal and unpigmented cells, has originated in the buds formed at the junction points of two grafted plants.

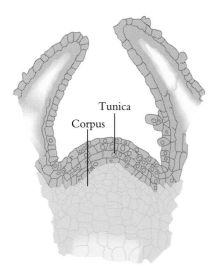

The tunica and corpus generative layers in the stem apex of *Coleus blumei.*

Shoot Development versus Root Development

When a shoot apex produces new cells, some usually differentiate to form organs, such as the leaves on a stem, that are spatially separated in a predictably ordered pattern that depends on the species. The pattern produced in any one case is believed to arise through the cooperative and sometimes competitive activities of a surface meristematic layer called the *tunica* and interior meristematic layers called the *corpus.* We may conceive of the tunica as giving rise to the surface of a new organ, while the corpus produces the bulk or interior tissue. If the tunica grows more actively than the corpus, the newly formed shoot will have various outpocketings on its surface, which might give rise to separate structures such as leaves or floral organs. If, however, the corpus is more active, then the apical meristem is more likely to produce a cylindrical tissue mass with relatively few outpocketings. Control of the comparative activities of the tunica and corpus may therefore account for many observed patterns of differentiation, although most researchers in plant development take a somewhat more complicated view.

Once a center of differentiated cells such as a tiny leaf has developed through the interactions of tunica and corpus, it exerts control over the developmental patterns of cells around it. The neighboring cells are transformed into leaf cells until the complete leaf is formed. For at least some distance around each formed organ, there is a zone in which no differentiation occurs: The cells remain uncommitted. This type of interaction helps to account for the patterns of phyllotaxy on stems, in which leaves tend to be formed according to standard spatial formulas such as five leaves for every two spiral traverses around the stem.

Root development proceeds quite differently. Branches that arise from roots are not produced at surfaces as in stems, but rather from deep within the tissue. Recall that branch roots arise from localized divisions of pericycle tissue lying just within the vascular cylinder of the root. The groups of cells produced by such activity organize into meristems, which then grow their way out through the cortex and penetrate the exterior of the root, producing an obvious branch root. Because of this difference in the site of origin of branch organs, roots are much more limited in the kinds of structures they can produce than are stems.

As one looks around the plant kingdom, one is struck by the tremendous diversity of reproductive modes and the variety of mechanisms employed by plants for regenerating damaged organs of various kinds. Ultimately the power of regeneration resides in each single cell or even, as we have seen, its protoplast. But it is also true that the fate of a regenerating tissue mass depends on its size, position, and cellular composition as well as on the nature of the surrounding cells. In the field of regeneration, plants are not limited to the dull monotony of a single mechanism.

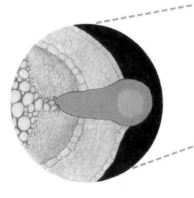

Microorganisms are found everywhere in the biosphere; indeed, these opportunistic organisms have colonized virtually every available ecological niche in nature. On the ocean floor, they make a bizarre living by oxidizing iron and sulfur released from volcanic vents; their ability to metabolize these elements has made them the ultimate source of nutrients and energy for an elaborate animal community. Deep within the soil, they use the energy gained by the oxidation of organic matter to secrete acids that degrade even feldspar and quartz, thus aiding in the formation of soil from rock. They range as high as the highest mountain tops, and their spores can be found in the upper reaches of the atmosphere. Because of their ubiquity, it is not surprising that they should frequently encounter higher plants, and because of their metabolic versatility, it is not surprising that they should interact extensively with them.

Much of the surface of the plant body is covered with a film of bacteria and other microorganisms. For example, the growing zone at the tip of the root synthesizes organic molecules of various kinds needed for root growth; some of these molecules leak or are excreted into the surrounding soil (the rhizosphere), where they attract motile bacteria and other microbes, including filamentous fungi. These organisms feed on the ex-

SEVEN

Cooperation
with Microbes

creted organic matter and in turn produce chemicals that can affect the green plant. There are fungi, for example, that release the hormones gibberellin or ethylene, which can stimulate or inhibit the growth of plants. Cytokinin synthesized by *Corynebacterium,* the organism producing witches'-broom, stimulates the growth of lateral buds, which normally remain dormant under the influence of apical dominance. Both free-living and symbiotic bacteria can help convert nitrogen gas from the atmosphere into ammonia in the soil, thus providing the plant with one of its essential elements.

Some microorganisms are able to colonize the plant either above or below the ground. Members of the genus *Agrobacterium* invade wounds in roots or stems, where they sometimes transform the growth of normal cells to produce abnormal structures like galls or masses of adventitious roots. Such transformed cells contain an altered form of the bacterial cell called a bacteroid, together with circular plasmids that transfer genes between the microbe and the higher plant. Because of its ability to transfer genes, *Agrobacterium* has found considerable use in genetic engineering operations.

The surface of the leaves may also be partly covered by colonies of microbes that affect the plant, sometimes in subtle ways. Recall that some bacteria of the genus *Pseudomonas* produce a protein that favors ice nucleation; in the presence of these bacteria, leaves may be damaged at otherwise benign temperatures. Still other leaf bacteria form dense colonies that completely block stomata, thereby impeding the entry of carbon dioxide, the release of oxygen, and the evaporation of water. Flower and fruit surfaces, and even the intercellular spaces of stems and roots, may also harbor populations of microbes whose chemical activities can affect the plant's physiological processes and growth. Some plants maintain internal microbial populations even after their surfaces have been sterilized for tissue culture. These microbes eventually emerge into the medium, thereby "contaminating" the culture. It is sometimes difficult to obtain plant tissue that is certifiably free of all microbes.

Mycorrhizae

A root tip that excretes large quantities of food molecules lures fungal hyphae to grow in its direction. When contact is made, the microorganism and plant frequently form a stable association for mutual benefit, or symbiosis, called a mycorrhiza. These mycorrhizae grow on the roots of many forest trees; the plant furnishes organic nutrition for the fungus, and the fungus acts like a mass of fine roots, absorbing and transporting

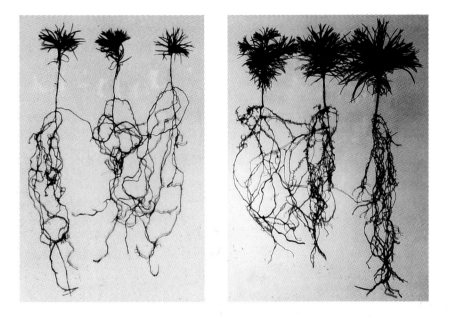

Nine months after germination, pine seedlings containing mycorrhizae (*right*) have grown far more vigorously than pine seedlings lacking the fungi (*left*).

mineral nutrients, especially phosphate, from the soil into the tree. Some trees growing on phosphate-poor soils will survive only if vigorous mycorrhizae have colonized almost the entire length of the root.

Some mycorrhizal fungi live mainly on the exterior surface of the root. Common in colder climates, these ectomycorrhizal fungi colonize

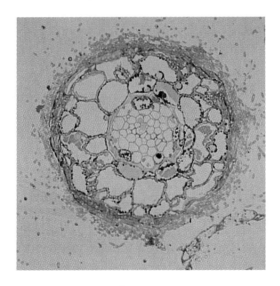

The rootlet of a pine tree covered by hyphae of an ectomycorrhizal fungus. Some hyphae penetrate into the intercellular spaces of the cortex.

the root tips of a range of host species, forming an external sheath of hyphal threads connected to an internal network of hyphae, mainly in the root cortex. The external threads feed and grow on organic molecules that are excreted by the root or are already present in the soil as decaying matter. The fungus obtains energy by oxidizing these food molecules, then uses part of this energy to absorb minerals and transport them to the interior hyphae, from which they make their way into root cells.

In most plants, however, the mycorrhizal fungi are entirely inside the plant body. Some of these endomycorrhizae, called VA (for *vesiculoar-buscular*) mycorrhizae, form special structures adapted for the storage of energy-rich lipids and for the interchange of nutrients between fungus and plant. Lipids are oxidized at the membranous junctions between plant and fungal cells to provide the energy for transporting materials across these membranes separating the fungus from the higher plant. In this instance, fungal morphology, rather than plant morphology, is altered by the association.

Orchids occupy a unique niche in the mycorrhizal world. In nature, orchids have an absolute requirement for the formation of mycorrhizae, since their seeds usually fail to germinate when the fungi are not present. If the seeds are infected by strands of the fungi *Rhizoctonia* or *Armillaria*,

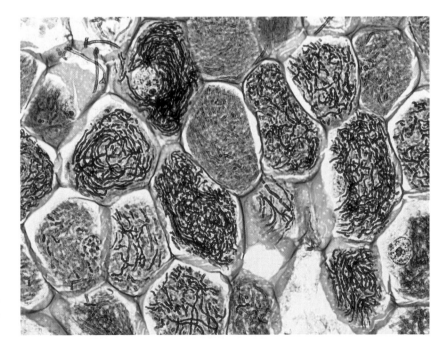

The hyphae of mycorrhizal fungi (dark) seem almost to fill some of the root cells of a host plant.

they germinate promptly. Oddly enough, these fungi typically act as pathogens on other plants, and they will assume the pathogenic habit on orchids as well if the latter are cultured on a medium that contains glucose. We have no idea why this should be so. Clearly, the fungi are temperamental members of this symbiotic association: they are cooperative when nutrients are in short supply and are harmful under other circumstances. Plant physiologists have learned to raise orchids without mycorrhizae by planting the orchid seed on especially rich sterile media. Thus, the fungi seem to supply the plant with otherwise unavailable nutrients: they may act by transforming the medium in some way and facilitating the entry of the altered products into the orchid body.

Nitrogen Fixation

Of the 17 elements known to be required by plants, 13 are derived from the weathering of rock particles, while only carbon, hydrogen, oxygen, and nitrogen are obtained from the atmosphere. Carbon, hydrogen, and oxygen enter the plant mainly in the form of carbon dioxide and water from the air and soil. Nitrogen exists in the atmosphere in its gaseous form, N_2, but in this form it is unavailable to higher plants and animals. Before these organisms can make use of the element, nitrogen gas in the atmosphere must be converted into ammonia (NH_3) in the process called nitrogen fixation.

In nature, nitrogen fixation is carried out mainly by varieties of bacteria. Some of these single-celled organisms, such as *Azotobacter, Azospirillum,* and *Clostridium,* may live free in the soil; most of these free-living bacteria obtain the energy for nitrogen fixation by oxidizing already formed decaying organic matter. Sometimes, however, photosynthesis powers nitrogen fixation directly. Nitrogen can be fixed by "blue-green" photosynthetic bacteria such as *Nostoc* and *Anabaena* or by other pigmented photosynthetic bacteria such as *Chromatium* (purple), *Chlorobium* (green), and *Rhodospirillum* (red). The different pigments of these bacteria act like chlorophyll: they capture light energy and transduce it into the chemical form of ATP. The ATP powers the synthesis of food molecules, whose energy is used to fix nitrogen. Because of their different colors, these pigments absorb different wavelengths of the spectrum to do their job.

The nitrogen-fixing, photosynthetic bacteria *Nostoc* and *Anabaena* commonly live in pockets in the fronds of the water fern *Azolla,* which grows abundantly in oriental rice paddies. Presumably the bacteria receive

Azolla filiculaides, the water fern, harbors nitrogen-fixing cyanobacteria like *Anabaena.* The ferns are surrounded by the small green plant duckweed *(Lemna).*

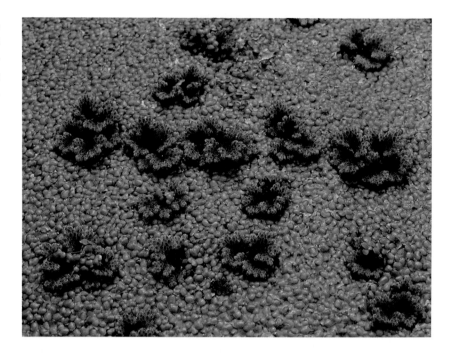

from the frond photosynthetically produced organic matter that helps their colonies to grow, while the fern receives fixed nitrogen in return. Although both *Nostoc* and *Anabaena* can live by themselves, they seem to flourish especially well in close association with *Azolla.* Peasants in the Orient have learned through experience that their rice crops flourish best when *Azolla* is first allowed to overgrow the paddy, then senesce and die. The death of the water fern releases various forms of fixed nitrogen into the paddy water; these are absorbed by the rice plants, which need ample nitrogen to yield a good crop of grain.

A nitrogen-fixing bacterium can form associations with many species of higher green plants; examples are found in the many symbioses between legumes and cells of *Rhizobium,* a bacterium that colonizes pea, bean, clover, and alfalfa plants, and of *Bradyrhizobium,* which prefers soybean plants. The nitrogen is fixed in special nodules formed when the bacteria invade the roots of the legume. Under normal conditions, neither the legume alone nor the free-living bacteria can fix nitrogen, yet when associated in the nodule they produce not only enough fixed nitrogen to maintain the growth of the nodulated plant, but enough to enrich the soil for the growth of neighboring plants.

The great majority of leguminous species form active nodules in response to only one strain of bacterium. There exist at least five groups of legumes, each responding best to a particular bacterial strain. Although other strains may be able to elicit nodule formation, their nodules are not effective fixers of nitrogen because some key component of the system is absent. Thus, in agricultural practice, unless the seeds or the field are inoculated with the appropriate organism, not enough nitrogen will be fixed to support a good crop yield.

Some nonleguminous plants fix nitrogen in nodules formed by association with microorganisms other than bacteria. The alder tree (*Alnus*) forms active nodules when infected by *Frankia,* a filamentous bacterium with rounded vesicles at its tips, in which nitrogen fixation presumably occurs. The same bacterium can nodulate a variety of other tree species belonging to unrelated families. This association thus seems less specific than the legume–*Rhizobium* symbiosis. Less stringent still is the association between the bacterium *Azospirillum lipoferum* and the roots of various tropical grasses, including corn. The growth of a corn plant associated with *Azospirillum* increases as the bacterial growth in the vicinity of the root tip becomes denser. It appears that *Azospirillum* cells do not form specialized structures, nor do they even penetrate the membrane of any plant cell; rather, they enter the root and grow in the intercellular spaces. This site of fixation outside the cell is in contrast to the leguminous nodule, in which nitrogen is fixed entirely within the cell.

The nitrogen-fixing nodules on bean roots formed by association with the bacterium *Rhizobium*.

The formation of nitrogen-fixing nodules in legumes resembles an elaborate ballet, in which each partner performs an intricate step, waits for a response, then builds on that sequence of events to move the dance along. Nodulation begins as motile bacteria are attracted to the surfaces of root hair cells by a "cocktail" of excreted substances, usually including sugars, amino acids, and phenols. The motile bacterial cells are somehow able to sense differences in the concentrations of these substances. Once they have encountered a few of the molecules, they swim toward the areas of higher concentration by adjusting the direction in which they beat their flagella, the whiplike fibers extending from their bodies. Such movement toward a chemical stimulus is called chemotaxis.

If the surface of a bacterium and a host cell are compatible, the bacterium recognizes and attaches to the cell surface. Specifically, a glycoprotein (a sugar attached to a protein) on the surface of a plant cell membrane binds with either a calcium-binding protein or a specific polysaccharide on the bacterial surface. Once the attachment is firm, the presence of the bacterium sets off new enzymatic processes in the plant cell, leading to the synthesis and excretion of a particular flavonoid (one of the same family of compounds that may be important in tendril coiling). The flavonoid is absorbed by the bacterial cell, where it is recognized by a protein whose synthesis is controlled by one of more than 40 types of nodulation (*nod*) genes. The coupling of flavonoid and protein activates other *nod* genes to produce specific lipopolysaccharides, molecules that are partly lipid and partly carbohydrate.

The lipid part of this *nod* factor determines which plant cells will recognize and attach to the bacterium. The lipid has a carbon chain with a variable length and a variable number of carbon-to-carbon double bonds, depending on the strain of bacterium. Only those plant species within the bacterium's host range will recognize lipids of its chain length and number of double bonds. The carbohydrate-like portion of the molecule consists of 4 or 5 linked units called *N*-acetylglucosamine. These units are sugars to which an amino group (NH_2) and acetic acid are attached; they resemble the basic units of chitin, the substance of which insect skeletons are made. At the end of its carbohydrate-like chain, the *nod* factor contains a sulfur atom, in the form of a sulfate group (SO_4). No other naturally occurring molecule has exactly this structure.

When these complex *nod* factors are excreted from the bacteria, they are taken up by nearby plant cells, where, at very low concentrations, they cause profound changes, especially in root hairs. These hairs are tubular protrusions, up to a centimeter long, near the root tip; each hair is formed by the elongation of a single epidermal cell. In response to *nod* factors, the

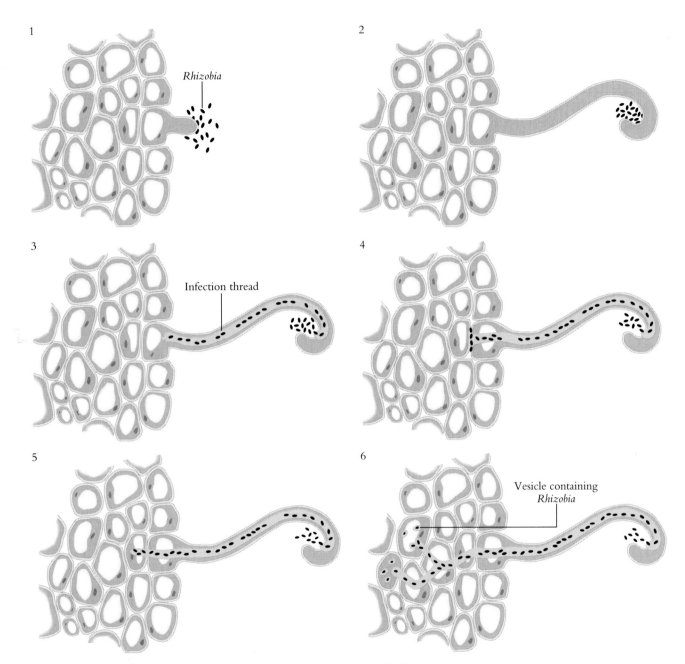

Early steps in nodulation. (1) Bacteria are attracted to a root hair. (2) The root hair curls. (3) Bacteria enter, and an infection thread forms. (4) The infection thread fuses with the membrane of the next cell. (5) The infection spreads to new cells. (6) Bacterial cells are released into the cytoplasm.

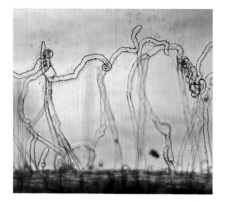

Alfalfa root hairs curl in response
to *Rhizobium.*

membrane of the root hair cell depolarizes, an indication that important changes in its permeability have occurred. Then, following marked changes in the internal cytoskeleton, the hair curls in a typical manner, forming a structure resembling a shepherd's crook. In this curled condition, the root hair is especially susceptible to invasion by the bacteria. At about the same time, cells in the root cortex several cell diameters away start dividing, probably in response to cytokinins released from the bacteria around the root hair. The new cells form either a discrete meristem or a dispersed zone of division. The body of the nodule will arise from repeated divisions of these activated cells.

Once the root hair has curled, bacteria enter the single cell of the hair, and some of the bacteria enlarge to become membrane-bounded bacteroids, which are the effective nitrogen-fixing form of the bacterium. (Bacteroids cannot divide, so for the infection to spread some bacteria must remain untransformed.) The plant responds by forming an infection thread, a tube made of plasma membrane that grows inward from the surface of the infected root hair cell. Although the infection thread is created by the plant, it carries the bacteroids and bacteria. When the infection thread reaches a barrier posed by a cell wall, the microbes break through the wall and a new thread starts up in the new cell, thus propagating the infection. Ordinary diploid cells in the vicinity of the infection thread usually die, while some of the naturally occurring tetraploid cells are stimulated to divide and grow. The area of growing cells enlarges and eventually merges with the nodule meristem formed earlier to create the nodule. These nitrogen-fixing cells, mainly tetraploid, become packed with bacteroids and bacteria, which are transmitted to new cells as the cells are formed.

Rhizobium produces cytokinin in culture, and it is believed that the growth of tetraploid cells in the developing nodule is preferentially promoted by cytokinin released from the bacteria, acting in combination with auxin produced by plant cells. Some of the flavonoids produced by infected cells are known to inhibit polar auxin transport: by interrupting the normal downward flow of auxin, they would likely cause this substance to accumulate at the point of infection. It is thus probable that high levels of both auxin and cytokinin in infected cells promote their division and extension. The high levels of these two hormones probably explain why the cells divide more rapidly than the cells around them. Ultimately, their more rapid growth produces the nodule, whose diameter may greatly exceed that of the root. The nodule cells cooperate with the bacteroids within them to synthesize many of the molecules essential to nitrogen fixation, including the leghemoglobin and dinitrogenase molecules discussed on page 200.

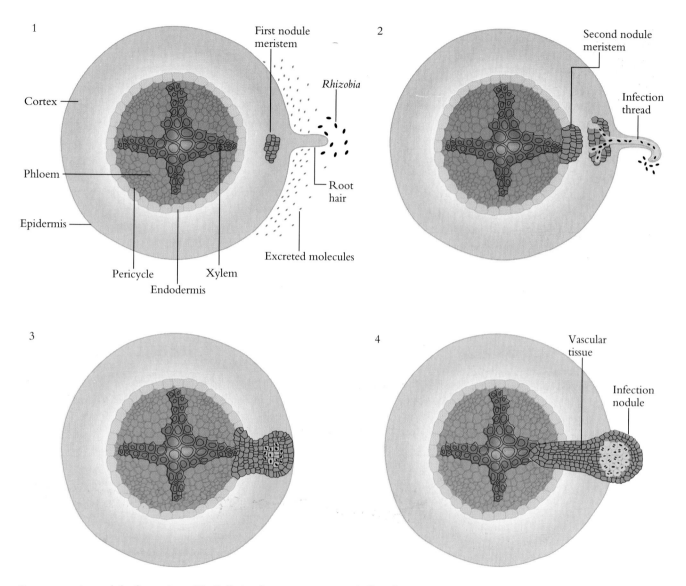

Later steps in nodule formation. (1) Cells in the cortex near an infected root hair start to divide. (2–4) The stimulus to divide eventually reaches the pericycle, and the emerging secondary root is altered to form a nodule.

Recently, it has been found that *Rhizobium* alone can fix nitrogen if it is grown in the virtual absence of oxygen. Oxygen is a potent inhibitor of nitrogen fixation, and part of the function of the nodule is to provide an anaerobic atmosphere within an aerobic cell for the functioning of the nitrogen-fixing apparatus. The oxygen level is kept low partly through

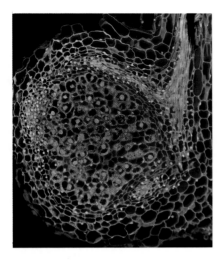

Dividing root cells have given rise to a nodule. The infected cells are those with red-orange cytoplasm and pink bacteroids. The small, red-orange cells at the tip of the nodule are meristematic tissue, while the blue and green tissues on either side of the nodule are vascular bundles.

the action of an iron-containing red pigment, leghemoglobin, which is related to the hemoglobin of animal blood. Like its animal analog, the leghemoglobin molecule combines with oxygen; in this way it lowers the concentration of the element in restricted regions of the cell. This pigment must be present if the nodule is to be active in nitrogen fixation, and in general, the redder the nodule, the more potent the nitrogen-fixing system.

The two nitrogen atoms of an N_2 molecule are held together by a relatively inert triple bond that is not easily broken. The bacteroid's primary task in fixing nitrogen is to break the bond and prepare the separated nitrogen atoms to accept three hydrogen atoms each. The hydrogen atoms are not attached whole, but as separated electrons and protons. The task has two steps: First, six electrons are transferred to the triple bond, with the expenditure of energy; a flow of six protons then follows to restore electrical neutrality. The cellular machinery necessary for nitrogen fixation is centered about the enzyme dinitrogenase, which transfers the electrons to the triple bond.

The dinitrogenase enzyme consists of two linked, iron-containing proteins. One of the proteins also contains two atoms of molybdenum (Mo); each of the atoms of molybdenum is in close association with an iron (Fe) atom to form a crucial Fe-Mo reaction center that binds to an N_2 molecule. It is at an Fe-Mo center that the N_2 receives the six electrons. We do not understand why the enzyme consists of two proteins rather than just one, but that situation is not uncommon. Metals like iron and molybdenum are good at moving electrons, which is probably what they do in nitrogenase.

The exact source of the electrons is also unknown, but most likely they are taken from organic acids formed during photosynthesis or during the breakdown of sugars in respiration; the organic acid pyruvate is a probable candidate. All organic acids contain a chain of carbon atoms terminating in a —COOH (carboxyl) group from which carbon dioxide is easily lost; in fact, the carbon dioxide liberated in respiration comes from such a decarboxylation reaction. As organic acids like pyruvate are oxidized and broken down during respiration, the energy liberated is conserved in the formation of ATP. Electrons are shuttled from the organic acid, one at a time, to the dinitrogenase protein that lacks molybdenum. From there the electron moves to the Fe-Mo reaction center in the second protein, where it is transferred to the N_2 molecule. Both proteins of dinitrogenase, but especially the first, are extremely sensitive to oxygen; the iron atom will bind to an atom of that element and lose the ability to accept electrons. For this reason, the proteins function properly only

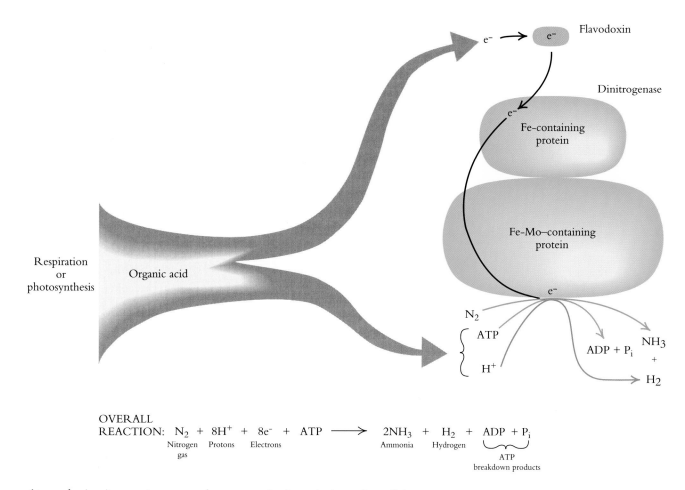

OVERALL
REACTION: N_2 + $8H^+$ + $8e^-$ + ATP $\longrightarrow$ $2NH_3$ + H_2 + ADP + P_i

Nitrogen   Protons   Electrons   Ammonia   Hydrogen

gas   ATP

breakdown products

Atmospheric nitrogen is converted to ammonia through the action of the enzyme dinitrogenase, which transfers electrons to a nitrogen molecule (N_2) so that it can accept protons (H^+) to become NH_3.

within the oxygen-free environment of leghemoglobin-rich compartments in the nodule.

When electrons leave the organic acid at the beginning of the process, they are carried to dinitrogenase by a molecule, probably flavodoxin, that acts as a transfer agent. This molecule is the enzyme that catalyzes the synthesis of ATP when pyruvate is broken down. The energy of the ATP molecules formed in this way is expended when the electrons are transferred to the Fe-Mo center; considerable energy, in the form of about five ATP molecules, is required for the transfer of each pair of electrons.

In the final step of nitrogen fixation, three protons are attached to each nitrogen atom to complete the synthesis of ammonia. These protons may be the products of photosynthesis or respiration. Alternatively, protons and electrons for the reduction of nitrogen to ammonia may be provided in the form of hydrogen atoms through the action of the enzyme hydrogenase, which splits molecules of hydrogen into individual atoms. Hydrogenase is found in certain microorganisms, but it is not generally found in higher plant cells. Its presence in the nodule is thus probably a result of transfer from the bacterium.

Once the ammonia is formed, it is released from the enzyme dinitrogenase and enters the mainstream of plant metabolism by attaching to an amino acid, glutamic acid, to form the related amino acid, glutamine. This reaction, catalyzed by the enzyme glutaminase, adds a second amino group ($-NH_2$) to a molecule already carrying one. In a related process called transamination, glutamine can pass this extra amino group to various organic acids, thereby converting them to amino acids. These amino acids are the 20 fundamental building blocks out of which all proteins are constructed.

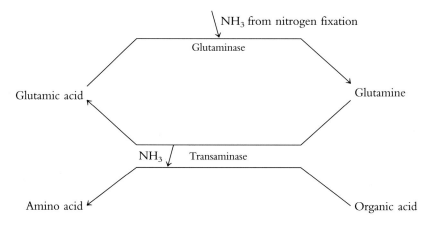

If the plant is fed any form of fixed nitrogen, especially ammonia or amino acids, both nodulation and nitrogen fixation are strongly inhibited. This makes good sense, since both processes are "expensive" from an energy point of view. Plants obviously prefer to take the easy road to obtaining their fixed nitrogen, by absorbing ready-made compounds if they are available, but they can switch to a more independent, energy-intensive mode when required. The detailed mechanisms of this response are still somewhat mysterious, but specific nodulation and nitrogen-fixing genes seem to be repressed by some proteins to which the ammonia or amino acid have become attached.

The Crown Gall Disease and Agrobacterium Tumefaciens

One of the most astounding interactions between plants and microbes is found in the crown gall disease, long known to botanists and gardeners. Plants stricken with the disease develop tumorlike masses called galls along the stem or root, generally near the crown (the junction of the stem and root near the soil surface) or at other locations at which mechanical stress and wounding are likely.

These tumors seem to grow like cancers, enlarging independently of the general growth pattern of the plant, and giving rise to other tumors at distant locations. The disease can be transferred from one locus to another, either on the same plant or on different plants, by transferring juice from the tumor into a previously wounded site. Thus, active tumors evidently contain an infectious agent. The infection is established when cells of *Agrobacterium* invade wounds in roots or stems made some hours earlier. The bacterium itself is not the infectious agent; rather, the disease is transmitted by a piece of DNA that moves from the bacterium to the plant cell. This is the only known example of DNA being transferred between organisms belonging to two different kingdoms, and it represents a kind of natural genetic engineering. The bacterium serves as the genetic engineer par excellence, providing the DNA to be transferred as well as the complete machinery for the transfer. Like nodulation, this remarkable process involves bacterial and plant cells in an intricate and repeated series of interactions.

The process of infection begins with the wounding of a plant cell, when the cell releases a mixture of sugars, amino acids, and sometimes phenolic substances that attracts motile bacteria to the wound. In response to the presence of numerous bacteria, the plant synthesizes phenols in sufficient quantity to activate certain genes in the bacteria. These genes control the attachment of the bacterium to the plant cell. As in nodulation, a polysaccharide-like component on the bacterial surface recognizes and attaches to a glycoprotein on the host cell. Simultaneously, other activated genes cause a piece of the bacterium's DNA to be excised and replicated. Once the attachment is firm, the excised and replicated piece of DNA can begin its transfer to the host cell.

Most of the bacterium's DNA is contained in a single large chromosome. The DNA to be transferred, called T-DNA, is not found on this chromosome, however; rather, it is part of a smaller circular piece of DNA separate from the chromosome, called a plasmid. This tumor-inducing ("Ti") plasmid is present in all virulent bacteria; while not essen-

A crown gall has formed on a stem of tobacco after a wound was inoculated with cells of *Agrobacterium tumefaciens*.

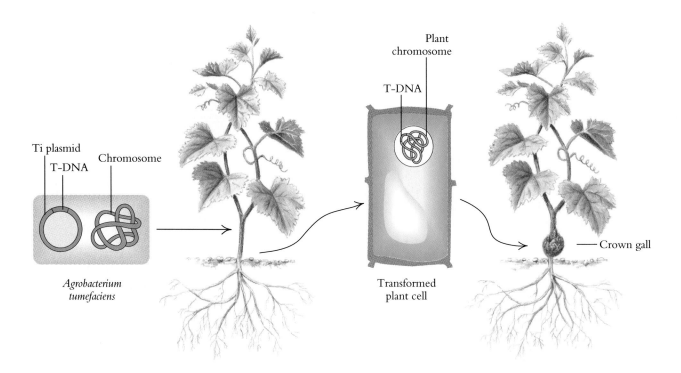

Agrobacterium tumefaciens induces a crown gall tumor by transferring a bacterial DNA from a plasmid to a plant chromosome.

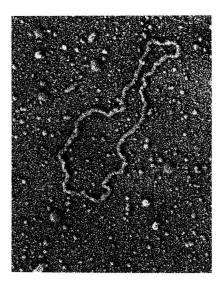

Electron micrograph of an *E. coli* plasmid used in DNA transfer and gene cloning.

tial for the life of the bacterium, its genes assist the microorganism's colonization of a foreign host. The plasmid contains virulence (*vir*) genes that, when activated by a suitable phenol from the plant, control the excision of T-DNA from the plasmid and its transport from the bacterium to the nucleus of the plant cell. Once the plasmid is inside the nucleus, additional *vir* genes act to incorporate the T-DNA into a plant chromosome, where it permanently transforms the plant cell.

The T-DNA contains genes that stimulate the synthesis of auxin and cytokinin. When these genes are "turned on" in the infected cell, its growth becomes autonomous: the cell continues dividing throughout the life of the plant, since it no longer depends on the transport of these hormones from distant loci. In this way, a tumorlike gall develops.

Another gene in plasmid DNA causes the infected cell to synthesize unique compounds called opines, which are formed by linking nitrogen-rich amino acids like arginine and lysine with common organic acids like pyruvate and ketoglutarate. The resulting complexes, rich in both carbon and nitrogen, cannot be metabolized by uninfected plant cells, but they can be broken down by infected cells containing bacteroids, altered forms of the invading bacterium, as in nitrogen fixation. Thus, the formation of opines guarantees the bacteroids a secure source of carbon and nitrogen. This aggressive process has been described as "chemical colonization." The opines also assist the transfer of the Ti plasmid from cell to cell and the activation of some of the virulence genes. As the original wound heals, and the concentration of phenol around the root accordingly declines, the rise in opine concentration continues the infection process, sometimes in cooperation with flavonoids manufactured by the plant.

T-DNA represents only about 10 percent of the total DNA of the Ti plasmid. Its borders are marked by special repeated sequences of base pairs that permit easy excision of the T-DNA. Any deletion, inversion, or other interference with the precise sequence of these 25 base pairs lessens or completely prevents the excision of the T-DNA. The T-DNA transferred from the bacterial plasmid to the plant chromosome is actually a copy of the portion of the plasmid containing the T-DNA, synthesized just before leaving the bacterium. When replicated, T-DNA appears as a single-strand copy, which is then combined with a specific protein for transfer to the plant nucleus. Were the T-DNA not protected by the protein, it would almost certainly be degraded by DNA-digesting enzymes in the plant cell. Such enzymes, called nucleases, are present in most cells to protect against the intrusion of alien DNA.

A variation of the crown gall disease is the "hairy-root disease," caused by a related bacterium, *Agrobacterium rhizogenes*. Instead of galls, plants infected with *A. rhizogenes* form masses of much-branched roots. Many of the molecular details of infection and colonization are similar to those of crown gall. One major difference is that *A. rhizogenes* activates the production of auxin much more than it does the production of cytokinin, thus accounting for the prolific growth of roots. Aseptic cultures of tissues transformed by the hairy-root disease organism have been used to study details of root metabolism and also to increase the yield of pharmacologically desirable products like alkaloids that are produced specifically in roots.

In a world populated by hordes of diverse microbes, the green plant has learned to co-opt some of them to help it survive in an often hostile environment. To achieve partnership with appropriate microbes, the plant has had to pay a price in energy expended. In return, it has gained access to scarce elements and genes helpful in its growth. Through its unique association with *Agrobacterium,* the plant has gained a genetic flexibility: the bacterium has probably introduced many useful genes into plants over the course of evolution. In the next chapter, we shall see how this flexibility has been put to work for the deliberate improvement of crop plants. Among these improvements is the introduction of genes to protect the plant against microbial pathogens.

A "hairy root" culture, formed by inoculating a root with *Agrobacterium rhizogenes*. Such cultures are useful in studying the biochemistry of roots and in studying products like alkaloids, formed in roots.

PHOTO RIGHT

The same gene can produce different effects when introduced into different genetic backgrounds. The maize pigmentation gene *R* promotes pigment formation in tobacco flowers (*upper right*) and hair formation on *Arabidopsis* stems (*lower right*). For comparison, unaltered plants are shown on the left.

DETAIL BELOW

The gene is transferred to the receptor plant by a bacterial plasmid, from which it makes its way to the plant cell nucleus.

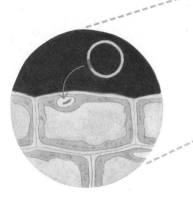

Ever since *Homo sapiens* first adopted an agricultural way of life, human beings have sought to improve on what they found in nature. Once they came to understand that plant characteristics are inherited, the most obvious technique was to choose good pollen for transfer to a desirable stigma, selecting only those individuals with the best attributes of hardiness, productivity, resistance to disease, and ease of cultivation. The genetically improved plant could be further helped by providing better nutrition, either in the form of a crude organic addendum like manure or decaying fish, or in the modern form of precisely formulated fertilizer. Finally, if the harvest was to be preserved from bacteria, fungi, insects, rats, or other predators, the plant needed the protection of physical barriers or chemical pesticides. Through selection and breeding, through nutritional fertilizing, and through the provision of protection from predators, humanity has vastly increased agricultural productivity, permitting the human population on earth to soar to more than 5.5 billion.

The modern corn plant, with its many rows of grain on large ears and its numerous large leaves on tall stalks, is a far cry from its primitive grasslike ancestors. Similarly, the modern large, red, juicy beefsteak tomato resembles only faintly the berrylike fruit of the small weed from which it was presumably derived. The time-honored practice of selection and breeding has indeed wrought miracles: by controlling plant reproduc-

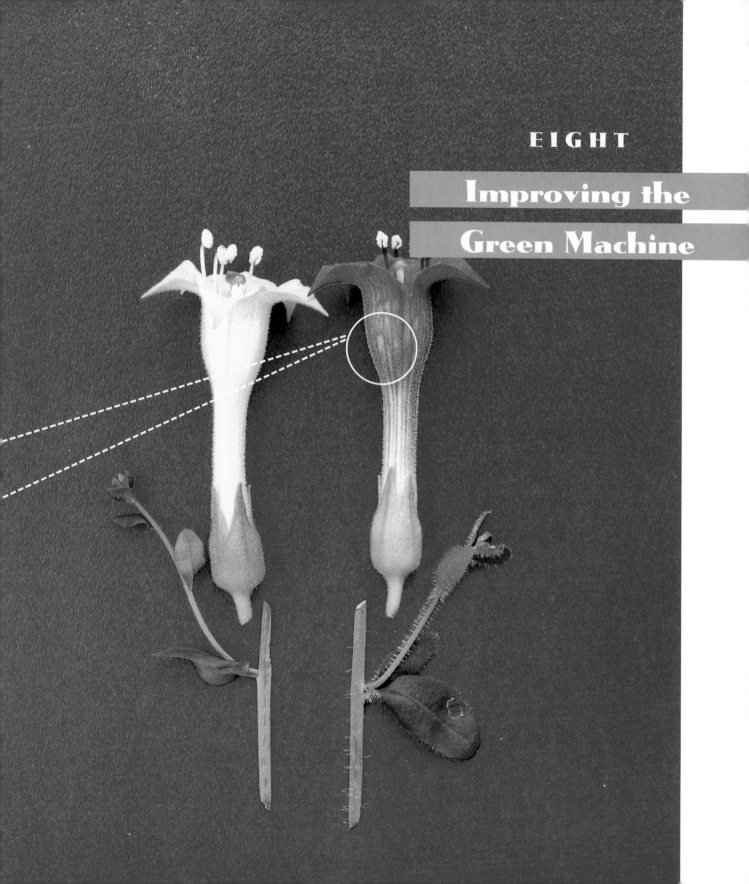

tion, human beings have been able to gather into single plant genomes those characteristics considered most desirable and to discard those considered unnecessary or harmful. However, neither modern corn nor the modern tomato could possibly survive in nature without human assistance.

Despite its successes, the traditional technique of selection and breeding suffers from serious theoretical limitations. The plant breeder has to work with what nature provides, and that does not always include the most desirable characteristics needed to improve the crop being bred. Sometimes the desired characteristic is entirely absent from plants, sometimes it is present in a different species of plant, but unavailable for transfer because the plant displaying the characteristic cannot produce viable offspring when crossed to the crop plant that one is trying to improve. Because of such problems, breeders have had to resort to various tricks. Most commonly, they have induced the appearance of new and useful variations by exposing seeds to irradiation or to chemicals that cause genetic mutations. From the mutants produced by such treatments, new variants can be selected for breeding purposes, and the desired genes incorporated into important older crop lines.

Another trick is to bypass natural barriers to interbreeding by fusing cells other than gametes. Through the fusion of protoplasts, some desired but previously unavailable crosses, blocked by incompatibility between the pollen and stigma, have been produced artificially. This technique has so far yielded no important new commercial varieties, but it is still in its infancy and may give better results after further exploration.

With some conventional crosses, the pollen and stigma are compatible, but the progeny are unable to progress beyond a certain developmental stage, usually in the embryo. Such developmental blocks can be bypassed and the progeny "rescued" by excising the blocked embryos from the seeds and growing them in tissue culture on especially rich medium. In other cases, a cross may yield hybrids that cannot reproduce normally and therefore cannot transmit their desirable characteristics by seeds. The difficulty arises because the two sets of chromosomes from different parents do not pair well, as they must when the haploid sex cells are produced by "reduction division" of the diploid genome. These hybrids can be rescued by the use of colchicine, which converts an unstable hybrid diploid into a stable tetraploid. As we have seen, colchicine, a poison derived from the crocus plant, interferes with cell division. At the beginning of cell division, each chromosome is transformed into two daughter chromosomes, both identical to the parent chromosome. Colchicine prevents the two newly formed daughter chromosomes from separating.

Although the origin of modern maize is still contended, most botanists believe it developed from a Mexican grass, teosinte. The earliest ears of corn were much smaller than their modern relatives.

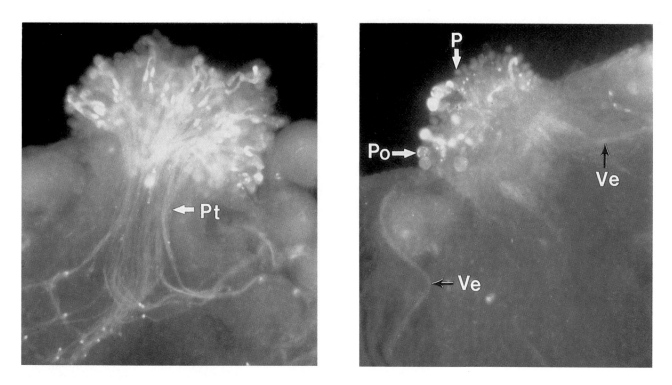

Breeding plants is tricky; a slight incompatibility between two plants will prevent their crossbreeding. When pollen lands on a compatible stigma (*left*), the tubes grow down the style to the ovary; on an incompatible stigma (*right*), pollen germination and growth are poor. P, papillar cells of stigma; Po, pollen grain; Pt, pollen tube; Ve, vascular element.

Both daughter chromosomes are therefore incorporated into the same cell, and since this happens to each chromosome in the diploid cell, the effect of colchicine is to double the chromosome count in each cell from diploid to tetraploid. Each chromosome now has a similar chromosome with which to pair; thus the tetraploid is stable, where the hybrid diploid was unstable. These tricks are a mere sampling of those available to the plant breeder; the exact technique used depends greatly on the problem presented and on the idiosyncracies of the plant.

Radical new approaches have become possible in recent years, since scientists have learned to regenerate entire plants from the protoplasts or isolated cells of certain species. For example, foreign DNA can now be introduced into and expressed in a protoplast. After the altered protoplast

In the *ablated* mutant of *Brassica* (*top photo, right*), few seeds are produced. The mutant stigma is malformed (*lower left*), and the mutant anther produces little viable pollen (*lower right*). A normal flower is shown in the top photo (*left*). W, wild type, normal; DTA, ablated mutant; An, anther; St, stigma.

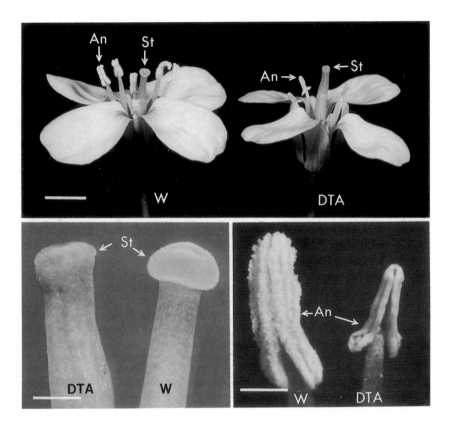

regenerates its wall and divides repeatedly to produce a callus and ultimately an entire plant, each cell of that plant will contain the newly introduced gene. To accomplish this kind of transformation, one first extracts DNA from a potential donor plant, using techniques that minimize DNA breakdown. This DNA represents many copies of the donor's complete genome. Once isolated in the lab, the donor DNA is introduced into protoplasts by one of a variety of fairly simple techniques.

Perhaps the most popular of these techniques is electroporation. Protoplasts are prepared in bulk to serve as receptors for the DNA and are then suspended in an appropriate medium containing the extracted donor DNA. After brief exposure to a mild alternating electric current, the medium and its contents are subjected to a momentary high-voltage shock of direct current. If the conditions are correct, the membranes of the protoplasts will "melt" and form temporary open pores at particular loci; these pores permit the entry of large molecules, including the foreign DNA. The pores then close, the membrane heals, and the protoplast goes on to develop, with some of the foreign DNA now inside the membrane. Under some circumstances, this foreign DNA is expressed spontaneously in the receptor protoplast; at other times special "promoter" DNA must be introduced to turn on the introduced genes. In both instances, the cellular machinery of the protoplast synthesizes the proteins encoded by the introduced DNA, and, if all goes well, the cell shows characteristics coded for by at least some genes in the foreign DNA. All of the cells of a plant regenerated from the protoplast should contain copies of the foreign DNA originally introduced by electroporation.

Other methods of introducing DNA are equally ingenious. Foreign DNA can be encapsulated in liposomes, microscopic spheres of natural or artificial lipids. These liposomes fuse with the natural membranes of protoplasts and deposit their DNA inside the cell. As with electroporation, the introduced genes can now become active. Perhaps the crudest yet most remarkable technique is the use of the "gene gun." By the sudden release of compressed gas, or by mild explosive charge, this device fires metallic pellets coated with DNA into intact plants or into discs punched from leaves. The force of the explosive charge is adjusted so that the pellet has enough momentum to penetrate into the tissue, but not enough to exit from it. Thus, the DNA is deposited somewhere in the cell; in a few cases it lands right in the nucleus and becomes fully functional.

Because it is impossible to predict where in the target the DNA will "hit," some method of visualizing the location of a successful entry is necessary. The usual solution is to splice "reporter genes" to the DNA being introduced. These genes either code for enzymes that produce colored products from chemicals supplied to the tissue or give some other obvious signal of their presence. For example, the GUS reporter gene produces glucuronidase, an enzyme that generates blue color wherever it acts, whereas the *lux* reporter gene encodes the firefly enzyme luciferase, which causes the cell to glow in the dark. The *cat* and *kan* genes confer resistance to the antibiotics chloramphenicol and kanamycin, respectively; their possessor will survive in the presence of the relevant antibiotic while other plants die. Whatever the nature of the reporter gene, its presence indicates the presence of the other genes to which it has been spliced.

Through any one of these physical methods, or through the use of a biological agent such as a virus or a plasmid, entire genomes or pieces of DNA can be transferred from one plant to another. But moving an entire genome from donor to receptor has an important limitation: potentially undesirable genes may be transferred along with the desired ones. Breeders have frequently pointed out how much better it would be if they could pluck out the single gene they wanted to transfer, place it in a

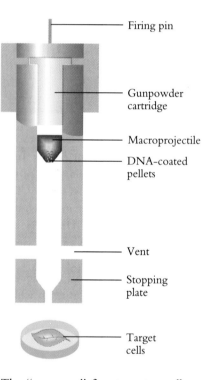

Firing pin

Gunpowder cartridge

Macroprojectile

DNA-coated pellets

Vent

Stopping plate

Target cells

The "gene gun" fires tungsten pellets coated with DNA into tissue and randomly transfers some genes to cell nuclei, where they may become active. A gunpowder charge accelerates the plastic macroprojectile holding the pellets. The macroprojectile stops at the plate, while momentum carries the pellets into the target.

plasmid or a virus that can enter the bacteria naturally. Such an entity, whose sole purpose is to introduce a piece of DNA into a new cell, is called a vector. The vector DNA is first cut open by the same restriction enzyme that was used to chop up the donor DNA; the cut vector DNA will thus have the same kinds of sticky ends as the donor DNA fragments. Then, each restriction fragment is mixed with the similarly cut DNA from the vector; under appropriate conditions, and with the help of enzymes called ligases, the two fragments fuse to form a stable circle of DNA, able to penetrate into the bacterium. Once inside, the introduced DNA is replicated whenever the bacterium divides. In this way, a single molecule of foreign DNA is converted to a "bucketful" of identical molecules.

Next, the bacteria are diluted and single cells are allowed to grow on the surface of a semisolid nutrient solution in a petri dish. Each cell produces a distinct colony. The descendants of each bacterium contain multiple copies of the cut piece of plant DNA carried by the bacterium; thus, the total library of DNA is contained in the total assembly of bacteria. But how does one find which colony of bacteria has the correct "book"—the piece of DNA containing the desired gene? It is rather like looking for a needle in a haystack, and any little trick that can narrow the search helps. One useful maneuver is to include in the vector DNA a gene conferring resistance to an antibiotic, such as chloramphenicol or kanamycin, and to include this antibiotic in the medium containing the bacteria. Thus, any bacterium that has picked up a circle of foreign DNA will be resistant to the antibiotic and will grow; any bacterium that has *not* picked up foreign DNA will be killed. This tactic reduces the number of cells to be examined, but still leaves the problem of finding the bacterial colony carrying the desired gene.

To find out which colony contains the gene for tomato worm resistance, one makes "replicas" of the pattern of the newly formed colonies by pressing moistened filter paper sheets over the bacterial cells in the petri dish. These replicas contain a sample of the cells from each colony. When removed and partially dried, they can be "probed" with appropriate substances, usually radioactively labeled antibody or pieces of DNA that will bind, respectively, only with the protein product of the desired gene or with the DNA of the gene itself. These probes come from a variety of sources. For example, we may already know that the resistant donor strain of tomato owes its resistance to the formation of the particular protein that is toxic to the tomato worm. We can therefore isolate quantities of this protein from the resistant donor plants and inoculate it into laboratory mice or rats to induce the formation of specific antibodies.

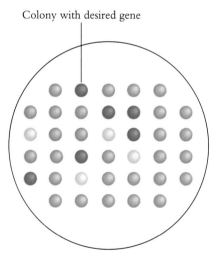

Colony with desired gene

Isolated bacterial colonies containing cloned DNA are incubated with a radioactive probe that is complementary to the desired gene. Hybridization of the two "lights up" the gene by producing an image on X-ray film.

Such antibodies, when extracted, purified, and made radioactive, will bind specifically with the protein produced by the gene, thus "lighting up" the colony containing the desired piece of DNA. (The protein that binds the antibody can only be produced by cloned genes in a special "expression" library.)

Locating the correct restriction fragment is not the end of the process, for the desired gene may be only a small portion of that fragment. To locate the resistance gene within the restriction fragment, we cut the DNA into even smaller pieces, using new restriction enzymes that cut at new locations. The entire process detailed above is repeated until finally

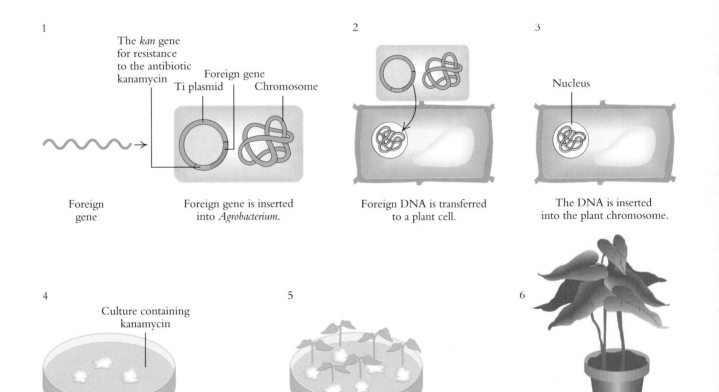

1

The *kan* gene for resistance to the antibiotic kanamycin
Foreign gene
Ti plasmid
Chromosome

Foreign gene

Foreign gene is inserted into *Agrobacterium*.

2

Foreign DNA is transferred to a plant cell.

3

Nucleus

The DNA is inserted into the plant chromosome.

4

Culture containing kanamycin

Only cells with the *kan* gene survive and divide in culture.

5

Cells regenerate into transgenic plantlets.

6

Plantlets grow into plants with new traits.

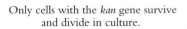

Transgenic plants are made by transfer of vectors containing foreign genes. Linking the desired new gene to the *kan* gene kills all nontransformed plants on a medium containing the antibiotic kanamycin.

the gene itself can be isolated. The gene is then cloned in the same way that the restriction fragments were cloned—by introducing into a vector suitably opened by a restriction enzyme a piece of DNA containing the gene and the requisite "sticky ends" to fuse with the opened vector. The fused and closed vector is now introduced into *E. coli* for cloning in the usual fashion. The most frequently used vector for the introduction of cloned genes into plant cells is the plasmid of *Agrobacterium tumefaciens;* some use has also been made of the cauliflower mosaic virus, which is capable of infecting many plants.

Finally, to indicate when a gene transfer into a plant has been successful, a reporter gene can also be included in the vector. In the absence of such a reporter gene, it is somewhat more difficult to tell if the desired gene is present in the regenerated plant, but its presence can usually be verified by observing the characteristics of the presumed transgenic plant. In our particular case, only a successfully transformed plant will be resistant to the tomato worm. But since this characteristic will obviously take considerable time to test, it is more desirable to insert a reporter gene like GUS, whose presence will allow a convenient molecular test to be performed immediately.

A somewhat more direct approach to isolating the resistance gene is to isolate the resistance protein for which it codes and determine its amino acid sequence. To sequence a protein, a single amino acid is removed from one end of an immobilized protein and identified by appropriate techniques. This step is repeated many times until the sequence of all amino acids in the protein has been determined, in order, from one end of the molecule to the other. Since amino acids in a protein are specified by triplets of bases in a nucleic acid, once we know the amino acid sequence we can deduce the possible sequence of bases in the nucleic acid of the gene coding for that protein. Then, using a commonly available laboratory device, "the gene machine," we can synthesize that nucleic acid directly, incorporate it into a plasmid, and clone it in bacteria for subsequent introduction into a higher plant. In addition, for convenience in subsequent detection of this synthesized gene, we sometimes introduce radioactive phosphorus (^{32}P) into its component nucleotides. Such a radioactive gene can be used as a probe for detecting similar genes in other plant species. When placed in a test tube with foreign DNA, it binds with stretches of DNA having a similar sequence, and the radioactivity signals the location of the piece of DNA with which it has bound. This technique may help isolate related resistance genes from other plants.

The flexibility of transgenic techniques is so great that the techniques have been used to introduce into plants genes from organisms of other

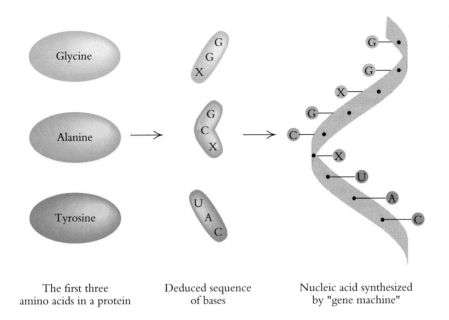

The first three amino acids in a protein

Deduced sequence of bases

Nucleic acid synthesized by "gene machine"

If we know the amino acid sequence of a protein, we can deduce the nucleic acid sequence of the gene controlling its production, then synthesize the gene in the laboratory.

kingdoms. Genes from bacteria have already proven helpful to plants. For example, the bacterium *Bacillus thuringiensis* ("*Bt*") produces a protein in its spores that is highly toxic to the caterpillars of certain butterflies and moths. The protein is thus potentially useful as a plant protectant. The effective *Bt* gene has in fact been isolated, cloned, and introduced into plants, where it has shown some success in conferring protection against insect predators. In tobacco, for example, *Bt* apparently substitutes for chemical insecticides fairly effectively. However, there is reluctance to introduce such a toxin into food plants, since it might turn out to be harmful to humans as well. For this reason, the *Bt* gene has thus far been inserted only into cells of trees and other crops not used as food. Even in such plants, the usefulness of *Bt* is threatened by the incredibly rapid rate of insect mutation. *Bt* has been in use for less than a decade, but already strains resistant to the *Bt* toxin have evolved. Nonetheless, similar toxins may prove to be valuable protectors against insects following the insertion of their genes into crops.

Genes were at first moved easily only into tobacco and other members of the *Solanaceae* family, but "gene jockeys" are gradually learning how to introduce them into members of the mustard family (*Cruciferae*) and other plant groups as well. Monocots have traditionally been more recalcitrant than dicots to the incorporation of foreign genes, but recently *Asparagus* and even some cereals have been successful recipients of transgenes. It appears that *Agrobacterium* plasmids containing a particular opine called nopaline can infect maize, while those containing the related opine octopine cannot. Other limitations and generalizations are constantly emerging from ongoing experimentation, and there now seems to be no theoretical limit to the manipulation of the plant genome. We are restricted only by the limits of our own ingenuity.

Academic investigators in this difficult field have learned to cooperate, freely sharing genetic libraries and probes, usually by mail. It is therefore not infrequent for collaborators listed on research publications never to have met one another! The absence of personal contact has raised trenchant questions about scientific accountability. What data are you responsible for if you are listed as an author on a research publication resulting from joint work with people you have never met? Where commercial organizations participate in research collaborations, additional complications are introduced. Since some transgenes are potentially quite valuable, commercially employed scientists may be reluctant to share procedural information, libraries, and probes until the appropriate patents have been obtained. This reluctance can sometimes cause problems when academic and industrial scientists convene at scientific meetings, since the former tend to be more open and expansive about their discoveries, while the latter may have to be more guarded.

Overripe wild type (*right*) and transformed (*left*) tomato fruits. The presence of a foreign gene that inhibits ethylene formation greatly lengthens the shelf life of the transgenic fruit.

Recently, large companies interested in agricultural biotechnology have given considerable sums of money to universities to fund research in areas of joint interest. In return, the university has sometimes promised to give the sponsoring company exclusive rights to develop any research discovery. Some academics have expressed concern over this warping of the traditional academic stance of open publication and free accessibility of research discoveries. Public research money was used on a large scale in the past to establish the basic principles of genetic technology that are now yielding potentially valuable data; the early role of public money would seem to compromise the right of private industry to obtain exclusive patent rights to some discoveries of major economic and practical importance. Moreover, when valuable genes, such as those for disease resistance, are obtained from plants of third world countries, should not these countries share in the massive profits that may flow from their exploitation in crops? At present, there is no formal arrangement for such sharing, although the 1992 International Biodiversity Treaty, if approved, will address the problem. At the very least, it appears to some that the engineered crop should be offered to the poorer country on exceptionally favorable terms, since that country would otherwise not be able to share in the rewards of the new discovery.

Mobile Genetic Elements

We have seen that scientists must first identify and isolate desirable genes before they can introduce them into the genome of a higher plant. One particularly promising technique for locating and isolating such genes makes use of spontaneously mobile genetic elements called transposons. These transposons can somehow move from one location to another; they are able to excise themselves from a DNA molecule and reinsert themselves at a new position, perhaps on a completely different chromosome. Transposons are able to move about the genome independently because their excision and insertion is performed by enzymes coded for by their own DNA.

The existence of these naturally "jumping genes" was deduced almost half a century ago by Barbara McClintock. While others worked with more "glamorous" organisms such as microbes and fruit flies, she worked alone and in virtual isolation with variously pigmented grains of Indian corn at the famous Cold Spring Harbor, New York, genetics laboratory. As she examined the patterns of pigmentation in grains on a single ear, she noticed many examples of reciprocal changes. A fully pigmented grain might be flanked by a totally unpigmented grain on one side and on the other side by a grain having pigment over only a part of its area. Yet all the cells in, for example, the endosperm of any given grain arise from the divisions of a single cell; should they not therefore all be alike?

After many years of detailed analysis, McClintock concluded that the pigmentation patterns were controlled by genetic elements that moved about throughout the genome. These elements were sections of the chromosomal DNA. Some of these elements had the power to excise them-

A gene *C* in these corn grains codes for the synthesis of purple pigment. If the mobile genetic element *Ds* (for *dissociation*) is inserted into this gene, pigment synthesis is turned off, and the grain will be all white instead of all purple. In the presence of another element, *Ac* (*activator*), *Ds* is transposed out of *C* in some cells, restoring pigmentation. If pigment is restored early in the development of the grain, the pigmented patch is large; if it is restored late in development, the patch is small.

selves from one location and insert themselves into another location; others controlled the movement of yet other dependent mobile elements. If a transposed genetic element inserted itself into the middle of another gene, such as a gene controlling pigmentation, it interrupted the coding of that gene: the gene was inactivated and no pigmentation appeared. But if it moved again, this time out of the cell whose pigment-producing system had been inactivated in this way, the pigmentation would be restored in the affected cell and all its progeny. If the ability to make pigment was restored to certain cells early in the development of the grain, they would divide many times before the grain reached maturity, and the pigmented section would be large; if the ability to make pigment was restored late in development, then the pigmented section would be small. For McClintock, looking closely at the ear's variously colored grains was like reading the history of the movement of transposons in cells.

McClintock's ideas were not greeted enthusiastically by geneticists, who had been brought up to believe that genes occupied fixed locations on the chromosome, like beads on a string. In fact, her work was either neglected or disbelieved for about 40 years, until microbial geneticists obtained independent evidence for the existence of transposons in bacteria. McClintock's work was "rediscovered" and its true significance appreciated. Well into her eighties, McClintock was awarded a Nobel prize and saw her work embraced and extended by an entire generation of new investigators. Several transposons have now been isolated and cloned, their structure determined, and the mechanisms of their excision, replication, and reinsertion defined. In addition, much of the chemistry of the steps in pigment formation and variation has been worked out. Thus, McClintock, who died recently at the age of 90, lived to witness the results of the revolution she had wrought.

Transposons are now widely used to induce mutations: they do so by inserting themselves into the middle of a gene, where they interrupt the gene's message and inactivate it. Since the experimenters who introduced the transposon know its structure, they can construct probes to easily locate where in the genome it has gone. They do so by making copies of the transposon that contain radioactive phosphorus and permitting these to "hybridize" with the transposon in a chromosome. Since complementary strands of DNA tend to find each other and stick tightly together, the transposon containing radioactive phosphorus sticks to the inserted transposon. Thus, finding the radioactive transposon simultaneously locates the inserted transposon and the inactivated gene. Then, by observing how the altered plant differs from a normal one, scientists can deduce the function of the inactivated gene.

The action of a transposon caused sex reversal in a usually male tassel. The normal plant (*top*) produces only male flowers in the tassel; the altered plant (*bottom*) produces female silks and ears in the tassel.

All plant transposons seem to be bounded at their ends by short, identical nucleotide sequences occurring in reverse order at the two ends; thus, one end might terminate in -AATTGC and the other in -CGTTAA. During the excision of a transposon, the enzyme transposase binds to these identical but reversed sequences and brings them together, forming a circular sequence that can be "lifted out" of the rest of the DNA. Sometimes the transposon excised in this way carries along neighboring bits of DNA when it moves, causing additional unexpected and unpredictable genetic upsets. Thus, cells subjected to "transposon mutagenesis" undergo major genetic alteration, as effectively as if they had been bombarded by radiation or treated with mutagenic chemicals. The advantage of using transposons to cause mutations is that one can locate the inserted transposon and thus the mutated gene.

In some plant species, transposons have been estimated to make up about 10 percent of the genome and to account for 50 percent of the spontaneous mutations. Thus, it is likely that transposon movement has been a major force in providing the variability essential to evolution. In addition, because the action of some genes is known to be affected by their neighbor genes, transposon insertion can produce aberrations in genes near but not at the site of insertion. For reasons we do not yet understand, transposons frequently remain quiescent and in place for a long time, but then undergo a burst of transposition activity. There is some evidence that such bursts are caused by severe environmental stress; if so, transposition may be called upon when needed to provide the new variants upon which survival depends.

Genetic Diagnostic Techniques

Many important crop plants chronically give low yields because they are systemically infected with symptomless viruses. Viruses are unable to reproduce without the assistance of a host; virus particles invade a plant cell and take over the cell's machinery for synthesizing nucleic acids and proteins, using that machinery to create new viral progeny. The synthesis of these viruses by the plant diverts energy and raw materials that could otherwise be used to produce new leaves, stems, and roots and to fill out seeds and ripening fruits. An instructive example is the barley yellow dwarf virus, which, although rarely noticed, is ubiquitous in cereal crops and may lower the yield by as much as 30 percent. Obviously, it would be advantageous to grow cereal crops without the virus. How can this be accomplished?

While growers may sometimes not be able to rid a crop of a systemic virus already present, they could avoid the use of virus-infected seeds and propagules if only they knew about their presence in advance. Here again, developments in molecular biology have come to their aid. Viruses can be spotted in a plant by appropriate nucleic acid probes or antibodies; when made radioactive or directly visible through a color reaction, these probes will "light up" those tissues that contain the viral component to which the probe binds. Such a test could be used to determine whether a cell or seed is virus-free, and thus safe for use in propagation.

Virus tests exploit the fact that a virus particle contains a protein coat and a nucleic acid core that can be easily separated by chemical means. The proteins in the coat can be purified and injected into laboratory mice or rats, which will produce specific antibodies to those proteins. Since those proteins are not normal plant components, but are produced only when the plant is infected by the virus, a reaction between antibody and

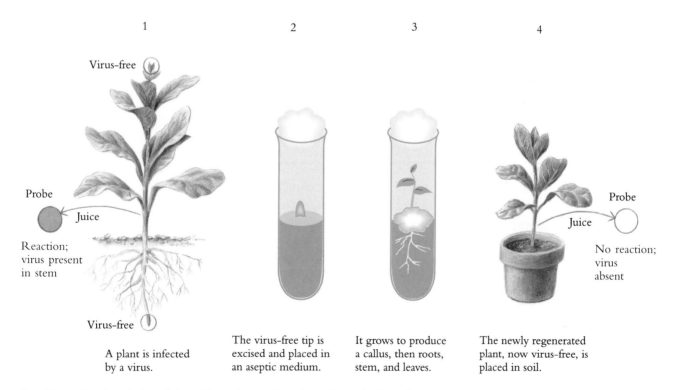

1
Virus-free

Probe
Juice

Reaction;
virus present
in stem

Virus-free

A plant is infected
by a virus.

2

The virus-free tip is
excised and placed in
an aseptic medium.

3

It grows to produce
a callus, then roots,
stem, and leaves.

4

Probe
Juice

No reaction;
virus
absent

The newly regenerated
plant, now virus-free, is
placed in soil.

Specific probes for viral nucleic acid can be used to determine whether plants contain or lack a particular virus. A virus-free plant can be regenerated from the virus-free root or stem apex of an infected plant.

plant tissue will indicate the presence of virus. If the antibody "probe" is made radioactive, or if it is coupled to a fluorescent component or to a color-producing enzyme, then a positive reaction between probe and plant can be easily detected. Similarly, viral nucleic acid made radioactive will hybridize with and "light up" virus-infected tissue. Commercial probes based on viral protein or nucleic acid are now widely used to test seed stocks and other propagules, and to certify certain plant materials as virus-free.

In some cases, the knowledge gained from these diagnostic techniques can be turned to therapy. Virus particles are large in comparison to the molecules that move through membranes; they are usually transported from cell to cell only through special strands of cytoplasm, called plasmodesmata. Since such strands usually develop only after a cell has enlarged and matured somewhat, they are generally absent from meristematic cells and the cells immediately behind them. But, as we have noted, small bits of apical tissue containing the meristem can be excised from the plant and cultured in artificial media. Under favorable conditions, such regenerated plants can be grown to maturity, and will produce flowers and viable seeds. Since such seeds are virus-free, they can be used to build up a stockpile of "clean" seed for farmers to sow. Over time, this new source of virus-free seedstock will also become infected, and the entire process must be repeated.

Although expensive, this new technique pays for itself in increased yield, and has become the basis for a new agricultural biotechnology based on the micropropagation of certified virus-free plant material. The technique is currently being used to derive high-yielding, certified virus-free strains of sugar cane from lower-yielding, infected strains. Since the virus-free strains may become reinfected after some time in the field, farmers must return to the tissue culture laboratories to buy probes to test their plants; if their plants are diseased, they must purchase new seeds periodically to replenish their stock. Although somewhat inconvenient for the farmer, the need to purchase reagents for probes and to restock at intervals makes micropagation a profitable venture for the laboratory. It also pays off for the farmer in increased yields.

Antisense Technology

Although all genes probably serve a beneficial function in the plant's natural environment, or they would not have evolved, the effects of some genes may be detrimental to the interests of the modern farmer or consumer. Thus, in some cases it may be as useful to remove or inactivate a gene as to insert one. As an example, some varieties of tomato have a short shelf life because they ripen rapidly under the influence of a massive production of ethylene. Were the production of ethylene to be halted, by removing or inactivating the gene that codes for an ethylene-producing enzyme, harvested tomatoes would be less likely to rot prematurely, and the consumer might enjoy cheaper and more abundant longer-lasting tomatoes.

Although particular genes can be inactivated by the insertion of transposons, or by exposure to radiation or mutagenic chemicals, these processes are random and inefficient, and are liable to harm other genes that we do not want to change. Specific genes can be inactivated more precisely by a method that makes use of so-called antisense technology. Recall that an active gene produces a protein, generally an enzyme that carries out a vital cellular activity. In this process, the two strands of the DNA helix separate, and one strand first makes a copy in RNA language. In this messenger RNA, the sequence of bases found in the DNA is preserved, except that the sequence is exactly complementary to the original pattern, since G will bind with C, and A will bind with T, and vice versa. The messenger RNA, now bearing the information of DNA in its own language, attaches to the ribosomes in the cytoplasm that synthesize proteins by "translating" RNA sequences into sequences of amino acids. If this is all done correctly, and the correct protein is produced, the message is said to exist in the correct *sense*. However, if an exactly complementary (*antisense*) RNA were present, it would combine with the single-stranded sense RNA and, thus it would annul the effect of the sense RNA message.

An antisense message can be introduced in one of two ways: an antisense RNA can be synthesized from the other DNA strand and introduced into a cell, or it can be transcribed directly in the cell by an introduced antisense gene. The sense RNA has to be in the single-stranded form to combine effectively with the ribosome; when specifically coupled to its own antisense RNA, it is blocked from further combination with the ribosome, and is inactive. Thus, the effect of the original gene is blocked specifically by its own antisense RNA, without direct effect on other genes. This technique has been used to control some important plant processes, including the ripening of tomatoes. The introduction of an antisense message for an ethylene-producing enzyme system annuls much of the ethylene production and lengthens the useful shelf life of the tomato. Because of its simplicity and specificity, this method has great promise for future agricultural use.

Introducing an antisense message for a membrane ATPase into a carrot plant (*right*) produces a stunted plant with a small root.

Ethical Considerations

Several environmentally concerned scientists, organizations, and lay persons have expressed concern over the release into nature of artificially produced genes or combinations of genes. Some militant groups have even disrupted properly designed and legally authorized tests of genetic engineering in agriculture. Citing the unexpected problems created by other chemical and physical intrusions into natural ecosystems, such groups question whether we may do more harm than good by our genetic manipulations. Just as we came to regret the addition of tetraethyl lead to gasoline because of the subsequent harm done to infant neural development by lead absorbed from the atmosphere, and just as we were forced to ban the use of DDT in fighting insects because of the terrible ecological damage it wrought on bird populations, these critics fear that new genetic systems released into nature may somehow become evil influences. They ask whether we can fully predict the consequences of deliberately releasing self-replicating genomes into nature. Couldn't a new gene endowing resistance to disease or herbicides escape from the targeted host plant and take up residence in some noxious weed whose growth would then create agricultural havoc? Couldn't some engineered sequence spliced into a plasmid or virus vector convert this vector into a noxious pathogen? Couldn't some new toxin against insects engineered into a crop plant turn out to be a subtle poison for humans as well? To these queries and conjectures, there can be no firm answers, only statements of probabilities.

When the potential dangers of recombinant DNA technology became apparent, several prominent scientists in the United States called for a voluntary moratorium on research in the field until its implications could be explored and some guidelines worked out. Following a historic conference at Asilomar, California, in 1975, the National Institutes of Health (NIH) set up guidelines for all of its grantees using recombinant DNA. Those who failed to adhere to these guidelines might find that their grants were cancelled, that they had earned the opprobrium of the scientific community, and that they were ineligible for future collaborative activities. Other government agencies have adopted similar guidelines, including the U.S. Department of Agriculture (USDA), which funds considerable plant research.

Recently, NIH has approved the use of recombinant DNA in the therapy of some human diseases. For example, a lack of the enzyme adenosine deaminase (ADA) in white blood cells can be fatal in humans. If, however, the white cells are removed from the body, genetically altered to contain genes for normal ADA, and then reintroduced into the body, the disease is "cured" in the treated individual, but not in that person's DNA. The therapeutic alteration of body cells other than sex cells is called *somatic* therapy, as opposed to *germ-line* therapy, which would affect the individual's reproductive DNA directly. So far, therapies have been allowed only if they cure a bodily illness by the genetic alteration of somatic cells that cannot be passed on to subsequent individuals through the germ line. Since the experiments with somatic treatments have been successful, further approvals for somatic therapies may be expected. Approval has not yet been given for experiments involving germ-line therapy, since there is still a disinclination to "play God" by meddling directly with human evolution.

Several genetic engineering experiments with food crop plants have been approved by the USDA, and the results have been encouraging. The gene controlling ice nucleation in bacteria living on the leaves of strawberries has been deleted, and the production of ethylene in ripening tomatoes has been halted by blocking a gene with an antisense message. The Food and Drug Administration (FDA) has ruled that foods will be judged on their quality and safety, not on the technique used to produce them. Thus, in accord with the opinions of almost all scientists, gene transfer by conventional breeding techniques and by genetic engineering are to be considered equivalent. (Notwithstanding, a group of gourmet chefs has recently been induced to oppose the use of genetically engineered foods in their preparations!) The federal regulatory agencies seem to feel less restraint in approving germ-line alteration of plant genomes. Some control is necessary, however, to prevent the malicious or accidental deterioration of food quality through gene transfer. Unfortunately, this type of operation is difficult to police, since the modest facilities of a genetic engineering laboratory could be set up secretly in a basement or garage. Those whose task it is to control research into potential biological warfare agents are especially concerned.

At present, NIH guidelines apply only to those receiving support from a federal agency. But what about commercial laboratories that do not require such support? Or wealthy and malicious individuals who can proceed on their own? The genetic engineering of plants and other organisms obviously has the potential to produce both tremendous human benefits and terrible evil. Regulatory agencies, legislation, and the development of ethical guidelines are likely to determine which of these predominates.

Plant Biotechnology and Third World Agriculture

Genetic engineering may in the future dramatically improve harvests, but development in this field requires the sophisticated use of advanced technology by highly trained scientists. Such research is thus not likely to be pursued vigorously in the lesser-developed countries of the world. For this reason, biotechnology may well increase the already considerable gap between rich and poor nations.

The question then becomes whether appropriate safeguards can be constructed to prevent that gap from widening. Nations themselves can obviously adopt some appropriate measures, such as mandating the sharing of royalties with other nations that have been the sources of plants, seeds, or genes used in profitable biotechnological advances. International organizations like the United Nations and foundations like the Rockefeller and Ford can establish training centers and programs to permit poorer countries to share in the newer technology. Fortunately, such actions have already been taken; for example, the International Rice Research Institute in the Philippines has already generated the "miracle rice" that sparked the Green Revolution, shared by many countries. However, these new varieties considerably outproduced conventional varieties only as long as the growing plants were supplied with large amounts of synthetic fertilizer and massively protected by chemical pesticides. The Green Revolution thus favored those who already had the capital to invest heavily in fertilizers, chemical pesticides, and mechanical devices, and was seen by some as exacerbating rather than resolving the productivity gap between the rich and poor nations.

The activity of international business organizations can also pose significant problems. For example, when the Monsanto chemical company discovered the effective herbicide glyphosate, it also found that the activity of this herbicide could be annulled by increasing the activity of a certain plant enzyme. Monsanto scientists then isolated the gene for this enzyme, cloned it, and introduced it into the seeds of crop plants. Now they could sell not only the herbicide, but also crop seeds resistant to the herbicide! Since third world countries probably cannot afford to buy both products, Monsanto's scientific success will probably further exacerbate the already large differences in agricultural productivity between the have and the have-not nations.

Should such inequities be allowed to develop naturally, or should Monsanto arrange to sell these products to lesser-developed countries at a much reduced price? Should an international agency like the World Bank help disadvantaged nations purchase such advanced technology, or is this unfair to the nations who have subsidized the education and science that have made such advances possible? As another example, the Unilever company in Britain has recently perfected the propagation of oil palm in tissue culture. If it exploits this discovery to the fullest, it might achieve a virtual global monopoly in the production of vegetable oil, and the sales of Philippine coconut oil and Senegalese peanut oil could well be damaged. Does the company have a responsibility to pay some restitution money to these relatively poor countries whose agriculture it may seriously damage?

Clearly, modern biotechnology cannot be divorced from the financial and political world that funds it. Since advances in science and technology cannot be stopped, it is essential that scientists cooperate with politicians and statesmen to formulate controls for the future. Plant science has matured, and the age of innocence is over.

Further Readings

General

Several recent books of botany and plant physiology explain the life processes of plants in terms reasonably comprehensible to the intelligent lay reader who has some knowledge of chemistry and physics. Among them are the following:

Galston, Arthur W., Peter J. Davies, and Ruth L. Satter. *The Life of the Green Plant,* 3d ed. Prentice-Hall. 1980.

Raven, Peter H., Ray F. Evert, and Susan E. Eichhorn. *Biology of Plants,* 5th ed. Worth. 1992.

Salisbury, Frank B., and Cleon W. Ross. *Plant Physiology,* 3d ed. Wadsworth. 1985.

Taiz, Lincoln, and Eduardo Zeiger. *Plant Physiology.* Benjamin/Cummings. 1991.

Wilkins, Malcolm B. (ed.). *Advanced Plant Physiology.* Pitman. 1984.

Wilkins, Malcolm B. *Plantwatching.* Roxby Reference Books. 1988.

Volumes of the *Annual Review of Plant Physiology and Plant Molecular Biology,* published yearly since 1950, carry advanced-level reviews of all subjects covered in this book and many more.

Prologue

Björn, Lars Olof. *Light and Life.* Hodder and Stoughton. 1976.

Esau, Katherine. *Anatomy of Seed Plants,* 2d ed. Wiley. 1977.

Fahn, Abraham. *Plant Anatomy,* 4th ed. Pergamon. 1990.

Margulis, Lynn. *Early Life.* Science Books International. 1982.

Schopf, J. William. *Earth's Earliest Atmosphere: Its Origin and Evolution.* Princeton. 1983.

Chapter 1

Bazzaz, Fakhri A., and Eric D. Fajer. Plant life in a CO_2-rich world. *Scientific American 266:* 68–74. 1992.

Clayton, Roderick K. *Photosynthesis: Physical Mechanisms and Chemical Patterns.* Cambridge. 1981.

Govindjee, and William J. Coleman. How plants make oxygen. *Scientific American 262:* 50–58. 1990.

Ledbetter, Myron C., and Keith R. Porter. *Introduction to the Fine Structure of Plant Cells.* Springer. 1970.

Miller, Kenneth R. The photosynthetic membrane. *Scientific American 244:* 102–113. 1979.

Rabinowich, Eugene I., and Govindjee. *Photosynthesis.* Wiley. 1969.

Youvan, Douglas C., and Barry L. Marrs. Molecular mechanisms of photosynthesis. *Scientific American 256:* 42–48. 1987.

Chapter 2

Bünning, Erwin. *The Physiological Clock.* Springer. 1973.

Hillman, William S. *The Physiology of Flowering.* Holt-Rinehart-Winston. 1962.

Kendrick, Richard E., and Barry Frankland. *Phytochrome and Plant Growth,* 2d ed. Edward Arnold. 1983.

Moses, Phyllis B., and Nam-hai Chua. Light switches for plant genes. *Scientific American 258:* 88–93. 1988.

Sage, Linda C. *Pigment of the Imagination: A History of Phytochrome Research*. Academic. 1992.

Smith, Harry. *Phytochrome and Photomorphogenesis*. McGraw-Hill. 1975.

Vince-Prue, Daphne. *Photoperiodism in Plants*. McGraw-Hill. 1975.

Chapter 3

Darwin, Charles, and Francis Darwin. *The Power of Movement in Plants*. Appleton-Century-Crofts. 1881.

Davies, Peter J. (ed.). *Plant Hormones and their Role in Plant Growth and Development*. Martinus Nijhoff. 1987.

Erickson, Ralph O., and Wendy K. Silk. The kinematics of plant growth. *Scientific American 242:* 134–151. 1980.

Schuck, Peter H. *Agent Orange on Trial: Mass Toxic Disasters in the Courts*. Belknap-Harvard. 1986.

Weaver, Robert J. *Plant Growth Substances in Agriculture*. Freeman. 1972.

Went, Frits W., and Kenneth V. Thimann. *Phytohormones*. Macmillan. 1938.

Chapter 4

Darwin, Charles. *The Movements and Habits of Climbing Plants*. John Murray. 1875.

Evans, Michael L., Randy Moore, and Karl H. Hasenstein. How plants respond to gravity. *Scientific American 255:* 100–107. 1986.

Gordon, Solon A., and Melvin J. Cohen (eds.). *Gravity and the Organism*. University of Chicago. 1971.

Hart, James W. *Plant Tropisms and Other Plant Movements*. Unwin Hyman. 1990.

Heslop-Harrison, Yolande. Carnivorous plants. *Scientific American 238:* 104–115. 1978.

Sweeney, Beatrice M. *Rhythmic Phenomena in Plants*. Academic. 1987.

Chapter 5

Cherry, Joe H. (ed.). *Environmental Stress in Plants*. Springer. 1989.

Key, Joe L., and Tsune Kosuge (eds.). *Cellular and Molecular Biology of Plant Stress*. Alan R. Liss. 1985.

Levitt, Jacob. *Responses of Plants to Environmental Stresses*, 2d ed. Academic. 1980.

Rosenthal, Gerald. The chemical defenses of higher plants. *Scientific American 254:* 94–99. 1986.

Turner, Neill C., and Paul J. Kramer (eds.). *Adaptation of Plants to Water and High Temperature Stress*. Wiley. 1980.

Chapter 6

Evans, David A., William R. Sharp, Phillip V. Ammirato, and Yasuyuki Yamada (eds.). *Handbook of Plant Tissue Culture. Vol. I: Techniques for Propagation and Breeding*. Macmillan. 1983.

Goldberg, Robert. Plants: Novel developmental processes. *Science 240:* 1460–1467. 1987.

Shepard, James F. The regeneration of potato plants from leaf cell protoplasts. *Scientific American 246:* 154–166. 1982.

Steeves, Taylor A., and Ian M. Sussex. *Patterns in Plant Development*. Cambridge. 1989.

Street, Herbert E. (ed.). *Plant Tissue and Cell Culture*. University of California. 1973.

Thorpe, Trevor A. (ed.). *Plant Tissue Culture: Methods and Applications in Agriculture*. Academic. 1981.

Wareing, Philip F., and I. D. J. Phillips. *The Control of Growth and Differentiation in Plants*, 2d ed. Pergamon. 1978.

Zimmerman, Richard H., Robert J. Griesbach, Freddi A. Hammerschlag, and Roger A. Lawson (eds.). *Tissue Culture as a Plant Production System for Horticultural Crops*. Martinus Nijhoff. 1986.

Chapter 7

Brill, Winston J. Agricultural microbiology. *Scientific American 245:* 198–215. 1981.

Chilton, Mary-Dell. A vector for introducing new genes into plants. *Scientific American 248:* 50–59. 1983.

Lumpkin, Thomas A., and Donald L. Plucknett. *Azolla as a Green Manure: Use and Management in Crop Production*. Westview. 1982.

Palacios, Rafael, Jaime Mora, and William E. Newton (eds.). *New Horizons in Nitrogen Fixation*. Kluwer. 1993.

Verma, Desh Pal S., and Normand Brisson (eds.). *Molecular Genetics of Plant-Microbe Interactions* (vol. 2). Kluwer. 1993.

Chapter 8

Beadle, George W. The ancestry of corn. *Scientific American 242:* 112–119. 1980.

Federoff. Nina V. Transposable genetic elements in maize. *Scientific American 250:* 84–98. 1984.

Gasser, Charles S., and Robert T. Fraley. Transgenic crops. *Scientific American 266:* 62–69. 1992.

National Research Council Staff. *Genetic Engineering of Plants: Agricultural Research Opportunities and Policy Concerns*. U.S. National Academy of Science Press. 1984.

Silver, Simon. *Biotechnology: Potentials and Limitations*. Springer. 1986.

Torrey, John G. The development of plant biotechnology. *American Scientist 73:* 354–363. 1985.

Sources of Illustrations

Plant renderings by Steve Buchanan; all other drawings by Hudson River Studio.

Page 1
Gerry Ellis

Page 4
Reproduced with permission from H. Friedli, H. Lötsher, H. Oeschger, U. Siegenthaler, and B. Stauffer, "Ice core record of the $^{13}C/^{12}C$ ratio of atmospheric CO_2 in the past two centuries," *Nature, 324:* 237–238. Copyright 1986 by Macmillan Magazines Limited.

Page 8
Holger Jannasch, Woods Hole Oceanographic Institution

Page 12
Ray Evert

Page 14
Peter Kresan

Page 16
William K. Sacco, collection of Arthur W. Galston

Page 19
Dwight Kuhn

Page 21
Adapted from Arthur W. Galston, *The Life of the Green Plant,* Prentice-Hall, 1961.

Page 25
Figure 8.C in Lincoln Taiz and Eduardo Zeiger, *Plant Physiology,* Benjamin/Cummings, 1991, p. 182.

Page 27
Lawrence Bogorad

Page 28
Adapted from Bruce Alberts et al., *Molecular Biology of the Cell,* 2d ed., Garland, 1989, p. 367, Figure 7-38.

Page 29
Adapted from A. Trebst, "The topology of the plastoquinine and herbicide binding peptides of photosystem II in the thylakoid membrane," *Z. Naturforsch, Teil C., 41:* 240–245 (1986).

Page 30 left
Runk & Schoenberger/Grant Heilman

Page 30 right
Alan Barkan, "Nuclear mutants of maize with defects in chloroplast polysome assembly have altered chloroplast RNA metabolism," *The Plant Cell, 5:* 389–402 (1993).

Page 38 bottom
Melvin Calvin

Page 40 bottom
Adapted from O. Bjorkman and S. B. Powles, "Inhibition of photosynthetic reactions under water stress: Interaction with light level," *Planta, 161:* 490–504 (1984).

Page 41
A. M. Clark, J. A. Verbeke, and H. J. Bohnert, "Epidermis-specific gene expression in *Pachyphytum*," *The Plant Cell, 4:* 1189–1198 (1992). Photos by Hans Bohnert

Page 42
S. E. Frederick, courtesy of E. H. Newcomb, University of Wisconsin/Biological Photo Service

Page 44
Adapted from J. A. Berry and J. S. Downton, "Environmental regulation of photosynthesis." In *Photosynthesis, Development, Carbon Metabolism, and Plant Productivity,* Govindjee, ed., vol. III, Academic Press, pp. 263–343.

Page 45
Douglas McCain et al., "Water is allocated differently to chloroplasts in sun and shade leaves," *Plant Physiology, 86:* 16–18 (1988).

Page 46
From Wolfgang Haupt, "Photomovement." In *Photomorphogenesis in Plants,* R. E. Kendrick and

G. H. M. Kronenberg, eds., Martinus Nijhoff, pp. 415–441.

Page 49
Rod Plank/Tony Stone Images

Page 52
Stephen Parker/Photo Researchers

Page 54
National Agricultural Library Archives. Photo courtesy of Linda C. Sage

Page 55
Adapted from Aubrey W. Naylor, "The control of flowering," *Scientific American, 227:* 49–56 (May 1952).

Page 57
Figure 5-12 in Arthur W. Galston, Peter J. Davies, and Ruth L. Satter, *The Life of the Green Plant,* 3d ed., Prentice-Hall, 1980.

Page 58
Laurie Campbell/Natural History Photo Agency

Page 60 top
From Harold McGee, *A Pigment of the Imagination: USDA and the Discovery of Photochrome,* U.S. Department of Agriculture.

Page 60 bottom
National Agricultural Library Archives. Photo courtesy of Linda Sage

Page 61
Adapted from Marion W. Parker, Sterling B. Hendricks, Harry A. Borthwick, and Norbert J. Scully, *Botanical Gazette, 108:* 1–26 (1946).

Page 62
From Lewis H. Flint and E. D. McAlister, *Smithsonian Institution Miscellaneous Colloquia, 96:* 1–8 (1937).

Page 63
Malcolm B. Wilkins, Botany Department, Glasgow University

Page 64
Malcolm B. Wilkins, Botany Department, Glasgow University

Page 65
From Harold W. Siegelman and Warren A. Butler, *Ann. Rev. Plant Physiol., 16:* 383–392 (1965).

Page 66–7
Frank B. Salisbury, Utah State University

Page 68
From E. Bünning, *Encyclopedia of Plant Physiology, 17(1):* 579–656 (1959), Springer-Verlag.

Page 70
After Georg Melchers and Anton Lang, *Biol. Zentr., 67:* 148 (1948).

Page 76
L. H. Pratt and A. Coleman, *American Journal of Botany, 61:* 195–202 (1974).

Page 77
Joanne Chory, *The Plant Cell, 3:* 445–459 (1991).

Page 78
Joanne Chory, *The Plant Cell, 3:* 445–459 (1991).

Page 79
Margaret Boylan and Peter Quail, *The Plant Cell, 1:* 765–773 (1989).

Page 81
J. Christopher Gaiser and Terri L. Lomax, "The altered gravitropic response of the lazy-2 mutant of tomato is phytochrome regulated," *Plant Physiology, 102:* 339–344. Photo by Christopher Gaiser

Page 82
Figures 6 and 7 in Charles Darwin, *The Power of Movement in Plants,* 1881.

Page 84
Malcolm B. Wilkins, Botany Department, Glasgow University

Page 89 top
M. Jacobs and S. F. Gilbert, *Science, 220:* 1297–1300 (1983).

Page 90 bottom
E. R. Degginger, Bruce Coleman Inc.

Page 91 top
From William A. Jensen and Frank B. Salisbury, *Botany,* 2d ed., Wadsworth, 1984.

Page 91 bottom
Ray Evert

Page 92 left
Charles H. Peterson, Institute of Marine Science, University of North Carolina

Page 92 right
Arthur H. Westing, Westing Associates

Page 95
B. G. Bowes, Botany Department, Glasgow University

Page 97
Jan D. Zeevaart, Michigan State University, East Lansing

Page 98
Sylvan H. Wittwer, Michigan State University

Page 100
Ian M. Sussex, University of California, Berkeley

Page 102
From Brian Capon, *Botany for Gardeners,* Timber Press, 1990.

Page 104
Adapted from Arthur W. Galston, Peter J. Davies, and Ruth L. Satter, *The Life of the Green Plant,* 3d ed., Prentice-Hall, 1980.

Page 105
Redrawn from Jean P. Nitsch, *American Journal of Botany,* 37: 211 (1950).

Page 106
Joseph R. Ecker, *The Plant Cell, 2:* 513–523 (1990).

Page 107
From Daphne J. Osborne, *Cotton Growing Review,* 51: 256–265 (1974).

Page 110
From Figure 32.17 in William K. Purves, Gordon H. Orians, and H. Craig Heller, *Life: The Science of Biology,* 3d ed., Sinauer Associates, W. H. Freeman and Company, 1992.

Page 113
Dwight Kuhn

Page 114
Nina Stromgen Allen, Department of Biology, Wake Forest University, and Marine Biological Laboratory

Page 119
From Arthur W. Galston, *American Scientist,* 55: 144–160 (1967).

Page 120
I. R. MacDonald et al., *Plant, Cell and Environment,* 5: 305 (1982). Photo by J. W. Hart, Department of Plant Science, University of Aberdeen

Page 121
Michael Evans, Ohio State University

Page 123
Randy Moore, Wright State University

Page 127
Stephen Parker/Photo Researchers

Page 129
From Mordecai J. Jaffe and Arthur W. Galston, *Plant Physiology 41:* 1014–1025 (1966).

Page 132
Adapted from Harry Smith, *Phytochrome and Photomorphogenesis,* McGraw-Hill, 1975, p. 164.

Page 133
Arthur W. Galston

Page 134
John Kaprelian/Photo Researchers

Page 136
Stephen Dalton/Natural History Photo Agency

Page 139
G. J. Cambridge/Natural History Photo Agency

Page 140
George Gainsburgh/Natural History Photo Agency

Page 143
Charles Gurche

Page 144
Adapted from T. Mansfield, *Journal of Biological Education, 5:* 115–123 (1971).

Page 146 left
Michele McCauley, collection of Ray Evert

Page 146 right
Ray Evert

Page 147
Based on data from Adrian K. Clarke and Christa Critchley, "The identification of a heat-shock protein complex in chloroplasts of barley leaves," *Plant Physiology, 100:* 2083 (1992).

Page 148
N. D. Hallam, courtesy of P. E. Kolattukudy, Institute of Biotechnology, Ohio State University

Page 152
Geoff du Feu/Planet Earth Pictures

Page 153
Laurie Campbell/Natural History Photo Agency

Page 156
Ciba Agricultural Biotechnology

Page 158
Eckhard Koch and Alan Slusarenko, "*Arabidopsis* is susceptible to infection by a downy mildew fungus," *The Plant Cell, 2:* 437–445 (1990).

Page 160
Peter Kresan

Page 163
Adapted from Hugh S. Mason and John E. Mullet, "Expression of two soybean vegetative storage protein genes during development in response to water deficit, wounding and jasmonic acid," *The Plant Cell, 2:* 569–579 (1990).

Page 164
Frank B. Salisbury, Utah State University

Page 167
William K. Sacco, collection of Arthur W. Galston

Page 172
Adapted from Guy Camus, *Rev. Cytol. et Biol. Véget., 11:* 119 (1949).

Page 174 bottom
Abraham D. Krikorian, Biology Department, State University of New York at Stoneybrook

Page 177 top
Edward C. Cocking, Nottingham University

Page 177 bottom
Barbara Schiller, R. G. Herrmann, and Georg Melchers, "Restriction endonuclease analysis of plastid DNA from tomato, potato and some of their somatic hybrids," *Mol. Gen. Genet., 186:* 453–459 (1982).

Page 178
James Carrington, *The Plant Cell, 2:* 987–998 (1990).

Page 180
Satish Maheshwari, Institute of Molecular Biology, Delhi University

Page 182
William K. Sacco, collection of Arthur W. Galston

Page 184
Jerome Wexler/Photo Researchers

Page 185
Ian M. Sussex, University of California, Berkeley

Page 186
From Peter H. Raven, Ray F. Evert, and Helena Curtis, *Biology of Plants,* 3d ed., Worth Publishers.

Page 189
Biophoto Associates/Photo Researchers

Page 191 top
Jaya G. Iyer, Department of Soil Science, University of Wisconsin, for the late Dr. S. A. Wilde, University of Wisconsin

Page 191 bottom
Robert D. Warmbrodt

Page 192
Runk & Schoenberger/Grant Heilman

Page 194
John Buckingham/Natural History Photo Agency

Page 195
Dwight Kuhn

Page 198
Mark E. Dudley and Sharon Long, "A non-nodulating alfalfa mutant displays neither root hair curling nor early cell division in response to *Rhizobium meliloti,*" *The Plant Cell, 1:* 65–72 (1989).

Page 200
Mark E. Dudley, Thomas W. Jacobs and Sharon Long, "Microscopic studies of cell divisions induced in alfalfa roots by *Rhizobium meliloti,*" *Planta, 171:* 289–301 (1987).

Page 203
Jo Handelsman and Steven A. Vicen, Department of Plant Pathology, University of Wisconsin-Madison.

Page 204 bottom
Stanley N. Cohen

Page 205
David Tepfer, CNRS, Versailles

Page 207
Cover of Science, 258 (December 11). © 1992 AAAS. Photo by Ted Preuss

Page 208
Dolores Piperno, Smithsonian Tropical Research Institute

Page 209
June Nasrallah, "*S* locus-directed cell killing in *Arabidopsis,*" *The Plant Cell, 5:* 253–261 (1993).

Page 210
Mikhail Nasrallah, "Ablation of papillar cell function in *Brassica,*" *The Plant Cell, 5:* 263–275 (1993).

Page 211
From Charles S. Gasser and Robert T. Fraley, "Transgenic crops," *Scientific American, 266(1):* 64 (June 1992).

Page 212
Robert Dietrich, Sharon E. Radke, and John J. Harada, "Downstream DNA sequences are required to activate a gene expressed in the root cortex of embryos and seedlings," *The Plant Cell, 4:* 1371–1382.

Page 218
Harry J. Klee et al., "Control of ethylene synthesis by expression of a bacterial enzyme in transgenic tomato plants," *The Plant Cell, 3:* 1187–1193 (1991).

Page 220
Nina Fedoroff, Carnegie Institute of Washington

Page 222
Erin E. Irish and Timothy M. Nelson, "Sex determination in monoecious and dioecious plants," *The Plant Cell, 1:* 737–744 (1989).

Page 226
J. Peter Gogarten et al., "The use of antisense mRNA to inhibit the tonoplast H^+ ATPase in carrot," *The Plant Cell, 4:* 851–864, (1992). Photo by Lincoln Taiz

Index

Other Books in the Scientific American Library Series

POWERS OF TEN
by Philip and Phylis Morrison and the Office of Charles and Ray Eames

HUMAN DIVERSITY
by Richard Lewontin

THE DISCOVERY OF SUBATOMIC PARTICLES
by Steven Weinberg

FOSSILS AND THE HISTORY OF LIFE
by George Gaylord Simpson

ON SIZE AND LIFE
by Thomas A. McMahon and John Tyler Bonner

FIRE
by John W. Lyons

SUN AND EARTH
by Herbert Friedman

ISLANDS
by H. William Menard

DRUGS AND THE BRAIN
by Solomon H. Snyder

THE TIMING OF BIOLOGICAL CLOCKS
by Arthur T. Winfree

EXTINCTION
by Steven M. Stanley

EYE, BRAIN, AND VISION
by David H. Hubel

THE SCIENCE OF STRUCTURES AND MATERIALS
by J. E. Gordon

SAND
by Raymond Siever

THE HONEY BEE
by James L. Gould and Carol Grant Gould

ANIMAL NAVIGATION
by Talbot H. Waterman

SLEEP
by J. Allan Hobson

FROM QUARKS TO THE COSMOS
by Leon M. Lederman and David N. Schramm

SEXUAL SELECTION
by James L. Gould and Carl Grant Gould

THE NEW ARCHAEOLOGY AND THE ANCIENT MAYA
by Jeremy A. Sabloff

A JOURNEY INTO GRAVITY AND SPACETIME
by John Archibald Wheeler

SIGNALS
by John R. Pierce and A. Michael Noll

BEYOND THE THIRD DIMENSION
by Thomas F. Banchoff

DISCOVERING ENZYMES
by David Dressler and Huntington Potter

THE SCIENCE OF WORDS
by George A. Miller

ATOMS, ELECTRONS, AND CHANGE
by P. W. Atkins

VIRUSES
by Arnold J. Levine

DIVERSITY AND THE TROPICAL RAIN FOREST
by John Terborgh

STARS
by James B. Kaler

EXPLORING BIOMECHANICS
by R. McNeill Alexander

CHEMICAL COMMUNICATION
by William C. Agosta

GENES AND THE BIOLOGY OF CANCER
by Harold Varmus and Robert A. Weinberg

SUPERCOMPUTING AND THE TRANSFORMATION OF SCIENCE
by William J. Kaufmann III and Larry L. Smarr

MOLECULES AND MENTAL ILLNESS
by Samuel H. Barondes

EXPLORING PLANETARY WORLDS
by David Morrison

EARTHQUAKES AND GEOLOGICAL DISCOVERY
by Bruce A. Bolt

THE EVOLVING COAST
by Richard A. Davis, Jr.

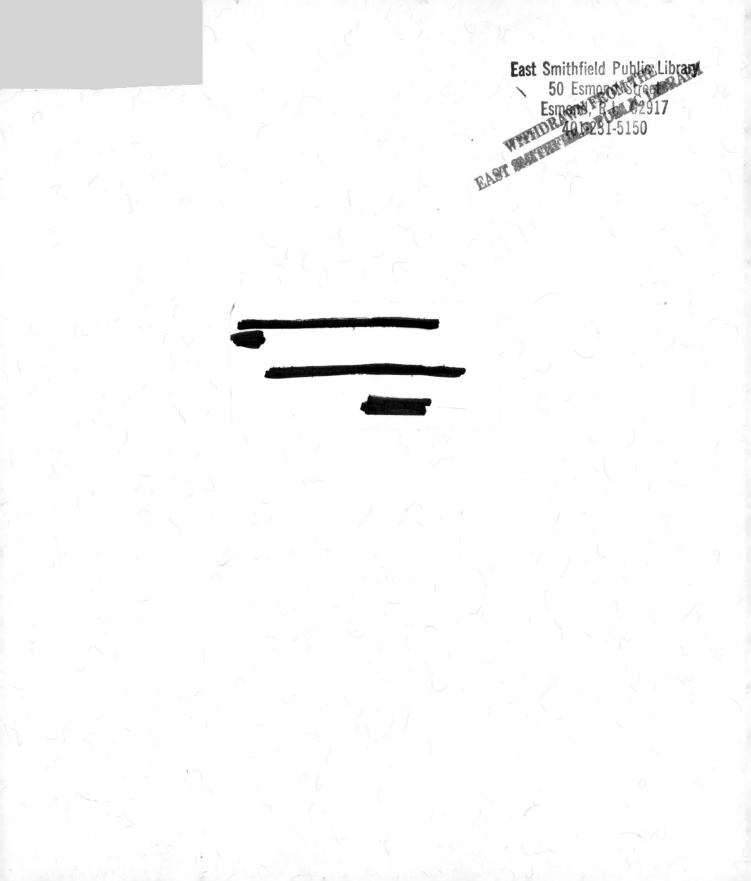